Physics and Chemistry in Space Vol. 16

Edited by L. J. Lanzerotti, Murray Hill and
D. Stöffler, Münster

A. Hasegawa T. Sato

Space Plasma Physics

1 Stationary Processes

With 45 Figures

Springer-Verlag
Berlin Heidelberg New York
London Paris Tokyo

Dr. Akira Hasegawa
AT&T Bell Laboratories
600 Mountain Avenue
Murray Hill, NJ 07974, USA

Prof. Dr. Tetsuya Sato
Hiroshima University
Institute for Fusion Theory
Higashi Sendamachi, Nakaku
Hiroshima 730, Japan

ISBN-13: 978-3-642-74187-6 e-ISBN-13: 978-3-642-74185-2
DOI: 10.1007/978-3-642-74185-2

Softcover reprint of the hardcover 1st edition 1989

Printing: Druckhaus Beltz, Hemsbach/Bergstr.
Binding: J. Schäffer GmbH & Co. KG., Grünstadt
2132/3145-543210 – Printed on acid-free paper

Preface

During the 30 years of space exploration, important discoveries in the near-earth environment such as the Van Allen belts, the plasmapause, the magnetotail and the bow shock, to name a few, have been made. Coupling between the solar wind and the magnetosphere and energy transfer processes between them are being identified. Space physics is clearly approaching a new era, where the emphasis is being shifted from discoveries to understanding. One way of identifying the new direction may be found in the recent contribution of atmospheric science and oceanography to the development of fluid dynamics. Hydrodynamics is a branch of classical physics in which important discoveries have been made in the era of Rayleigh, Taylor, Kelvin and Helmholtz. However, recent progress in global measurements using man-made satellites and in large scale computer simulations carried out by scientists in the fields of atmospheric science and oceanography have created new activities in hydrodynamics and produced important new discoveries, such as chaos and strange attractors, localized nonlinear vortices and solitons. As space physics approaches the new era, there should be no reason why space scientists cannot contribute, in a similar manner, to fundamental discoveries in plasma physics in the course of understanding dynamical processes in space plasmas. New satellite opportunities, which enable simultaneous observations of electromagnetic fields in a wide frequency range, and particles in a wide energy range, whereby the data can be correlated with those of other simultaneously flying satellites in order to provide space-time resolution will undoubtedly contribute to this new direction.

The Space Plasma Physics Volumes I and II are prepared to meet the demand for a theoretical basis for this new era of space physics. For this purpose, the book is arranged to incorporate with both plasma physics and space science with the help of computer simulations. The book is divided into two volumes. Volume I is designed to serve as introductory material for graduate students who are interested in exploring this field and for plasma physicists who wish to acquire experience in the physics of the space environment. Most of the materials which are covered in this volume are from standard plasma physics. Hence few references are cited in this volume.

The first two chapters of Volume I provide the necessary elementary knowledge of plasma physics to the graduate student who has no knowledge of plasma physics. Chapter 1 introduces the elementary methodology of plasma physics with a particular emphasis on the interplay between a plasma as a hydromagnetic fluid and as a group of discrete charged particles. Chapter 2 introduces small amplitude (linear) waves which are relevant in the space environment. It covers both kinetic and hydromagnetic waves in homogeneous as well as inhomogeneous plasmas.

Chapter 3 introduces stationary global plasma processes in space covering solar, interplanetary, and geomagnetic environments. The stationary processes provide a comprehensive preparation for understanding the complex dynamic processes in space which will be treated in Volume II.

MKS units are used throughout the text unless otherwise so specified.

Many colleagues of the authors have contributed to the preparation of this volume. Dr. L. J. Lanzerotti, the editor of this series, has encouraged us to undertake this adventure. Ms. C. G. Maclennan has helped us to improve the English presentation. Dr. K. Watanabe has provided us with many simulation results to make the book more colorful, and Dr. R. Horiuchi, Mr. C. Murakami and Y. Nakayama, our talented graduate students, spent considerable time reading the manuscript. Alyne Bonnell did an excellent job of typing the entire volume of this book. The authors are very grateful for their cooperation.

Since the author's residences are separated by some 150 degrees in longitude, correspondence over this distance has required generous support from their institutions, AT&T Bell Laboratories and Hiroshima University, as well as from their family members, Miyoko and Hisako. We would like to express our sincere appreciation to them.

Finally we would like to acknowledge the travel expense support by NSF grant ATM 86-09585 and NASA grant NAGW-894 for the authors' mutual visits to each others institutions, without which the completion of the book would not have been possible.

Akira Hasegawa and Tetsuya Sato, January 1988 at Hiroshima.

Table of Contents

Notations

(In the order of appearance in the text)

$\mathbf{E}(\mathbf{r},t)$:	Electric field vector
$\mathbf{F}(\mathbf{r},t)$:	Force field
q:	Charge
$\mathbf{v}(t)$:	Lagrangian velocity of a particle
$\mathbf{v}(\mathbf{r},t)$:	Velocity field
$\mathbf{v}$:	Independent variable representing the velocity coordinate in phase space.
$\mathbf{r}(t)$:	Lagrangian coordinate of a particle
ρ:	Charge density, Larmor radius
ϵ_o:	8.854×10^{-12}F/m, dielectric constant of vacuum
μ_o:	$4\pi\times10^{-7}$H/m, magnetic permeability of vacuum
$\mathbf{B}(\mathbf{r},t)$:	Magnetic field vector
$\mathbf{J}(\mathbf{r},t)$:	Current density
$\mathbf{P}(t)$:	Canonical variable
$\mathbf{Q}(t)$:	Canonical variable
$H(\mathbf{P},\mathbf{Q},t)$:	Hamiltonian
$\mathbf{A}(\mathbf{r},t)$:	Magnetic vector potential
$\phi(\mathbf{v},t)$:	Electric potential
$f(\mathbf{v},\mathbf{x},t)$:	Distribution function
ω_c:	Cyclotron (angular) frequency
subscript e:	Designation of electron
subscript i:	Designation of ion
m:	Mass
n:	Number density
ω_p:	Plasma (angular) frequency
v_T:	Thermal speed
T:	Temperature in energy unit
λ_D:	Debye wave (shielding) length
k_D:	Debye wave number
γ:	Collision frequency, adiabatic constant
λ_m:	Mean free path
$\ell n\Lambda$:	Coulomb logarithm, where $\Lambda = \dfrac{12\pi(\epsilon_o T/e^2)^{3/2}}{(n_e)^{1/2}}$
<>:	Average
E_K:	Kinetic energy density
$\mathbf{P}_K$:	Kinetic power flow

E_F:	Energy density of electromagnetic field
$\mathbf{P}_F$:	Power flow of electromagnetic field
$\mathbf{\Pi}_K$:	Kinetic momentum density
$\mathbf{\Pi}_F$:	Momentum density of electromagnetic field
$\mathbf{\Gamma}_K$:	Kinetic momentum density flow
$\overleftrightarrow{\mathbf{\Gamma}}_F$:	Momentum flow tensor of electromagnetic field
AU:	Astronomical unit (distance between the sun and the earth)
p:	Pressure
$\rho_s = (T_e/m_i)^{1/2}/\omega_{ci}$:	on Larmor radius at the electron temperature
μ:	Magnetic moment
$\mathbf{b}$:	$\mathbf{B}/B$; unit vector in the direction of the magnetic field
$w(=mv^2/2)$:	Kinetic energy
subscript $\perp$:	Components perpendicular to the magnetic field
subscript $\parallel$:	Component parallel to the magnetic field
ψ:	Magnetic flux density
$\mathbf{v}_E(=\mathbf{E}\times\mathbf{B}/B^2)$:	$\mathbf{E}\times\mathbf{B}$ drift velocity
$\mathbf{v}_p$:	Polarization drift velocity
$\mathbf{v}_R$:	Curvature drift velocity
$\mathbf{R}$:	Radius of curvature
$\mathbf{v}_B$:	Gradient B drift velocity
$\mathbf{J}_R$:	Curvature current
$\mathbf{J}_B$:	Gradient B current
$\mathbf{J}_p$:	Polarization current
$\mathbf{J}_M$:	Magnetization current
$\mathbf{J}_D$:	Diamagnetic current
Ω:	Vorticity
β:	Ratio of plasma internal energy density to the magnetic energy density
$\overleftrightarrow{\mathbf{p}}$:	Pressure tensor
ψ_D:	Magnetic flux function of the dipole field
J:	Action
ω_b:	Bounce (angular) frequency
J_2:	Second adiabatic invariant
η:	Resistivity
σ_e:	Electric conductivity
D:	Diffusion coefficient
$\mathbf{k}$:	Wave vector
k:	Wave number
λ:	Wavelength
ω:	Wave (angular) frequency

$\mathbf{v}_{ph}$:	Phase velocity
v_A:	Alfve´n speed
$\epsilon(\omega, \mathbf{k})$:	Plasma dielectric constant
$\mathbf{n}$:	Vector index of refraction
ω_{LH}:	Lower hybrid frequency
ω_{DH}:	Upper hybrid frequency
$c_s[=(T_e/m_i)^{1/2}]$:	Collisionless ion sound speed
Z:	Plasma dispersion function
ω_s:	Surface wave frequency.
σ:	5.67×10^{-8} joule/m^2K^4 sec, Stefan-Boltzmann constant
G:	6.67×10^{-11} m^3/kg sec^2, gravitational constant
$M_\odot$:	1.99×10^{30} kg, solar mass
κ_T:	Thermal conductivity
$\mathbf{v}_{gr}$:	Escape velocity
$\mathbf{v}_s$:	Solar wind velocity
M_E:	6×10^{24} kg, earth's mass
R_E:	6370 km, earth's radius
h:	6.6×10^{-34} joule sec, Plank's constant
σ_p:	Hall conductivity
$\sigma_{\parallel}$:	Parallel conductivity
μ_P:	Pedersen mobility
μ_H:	Hall mobility
$\mu_{\parallel}$:	Parallel mobility

Chapter 1. Physics of Stationary Plasmas

1.1 Introduction

This chapter is devoted to introducing the elementary physics of stationary plasmas in the space environment. Space plasma is basically tenuous, that is, the mean free path is generally comparable to or longer than the scale length of any inhomogeneity. Furthermore, the scale lengths of the inhomogeneities are often much larger than the characteristic scale length of plasma parameters such as the Debye length or Larmor radius (see Sect. 1.4 for the definitions). Under these circumstances, the motion of individual particles under the assumption of infinitely small Larmor radii (guiding center motion, see Sect. 1.7) is important in deciding how the plasma particles move in the space environment.

On the other hand, it is often the case in space plasmas that the plasma energy density is comparable to the energy density of the magnetic field. This fact suggests that tracing plasma particles under a prescribed "model" magnetic field can often lead to an erroneous result. Thus a "self-consistent" field which takes into account the magnetic field produced by the plasma current becomes important (see Sect. 1.10).

In order to obtain the plasma current, one must sum over the velocities of charged particles. When this is done for the guiding center positions, the magnetization current which originates from the Larmor motion of individual particles (see Sect. 1.8) must be added. This gives the total plasma current as the diamagnetic current, $J = \mathbf{B} \times \nabla p / B^2$, rather than the direct guiding center current. What is interesting is that this result is identical to that obtained from the magnetohydrodynamic (MHD) equations (Sect. 1.6) where the plasma is assumed to be a continuous fluid. This fact indicates the possible usefulness of the "fluid" approximation even for tenuous plasmas in the space environment. In fact, for many cases in which wave-particle interaction or heat conduction are not important, the MHD equations are useful because they are derived under the assumption that a Maxwellian distribution is maintained during the process of interest.

In dealing with space plasmas, a careful interplay between a fluid approach and a discrete particle approach is important.

This chapter is written with particular emphasis on these points. It is recommended that the reader digest carefully the contents of each section. In order to understand specific sections of interest in later chapters, the ideas presented in this chapter should be fully understood.

1.2 The Equation of Motion of a Charged Particle

In dealing with the dynamics of a plasma as a whole, one must first know the force acting upon each individual particle in the plasma and its resulting motion. In most cases in space plasmas, the dominant force is the one that originates from the electromagnetic field, the Lorentz force, $\mathbf{F}(\mathbf{r}, t)$.

$$\mathbf{F}(\mathbf{r}, t) = q[\mathbf{E}(\mathbf{r}, t) + \mathbf{v}(t) \times \mathbf{B}(\mathbf{r}, t)] \tag{1.2.1}$$

$$\mathbf{v}(t) = \frac{d\mathbf{r}}{dt} \tag{1.2.2}$$

and

$$m\frac{d\mathbf{v}}{dt} = \mathbf{F}(\mathbf{r}, t) . \tag{1.2.3}$$

Here q, m and $\mathbf{r}$ are the charge, the mass and the position of the particle. $\mathbf{E}$ and $\mathbf{B}$ are the electric and magnetic field vectors in MKS units which satisfy Maxwell's equations

$$\nabla \cdot \mathbf{E} = \rho/\epsilon_o \tag{1.2.4}$$

$$\nabla \times \mathbf{E} = -\frac{\partial \mathbf{B}}{\partial t} \tag{1.2.5}$$

$$\nabla \cdot \mathbf{B} = 0 \tag{1.2.6}$$

$$\nabla \times \mathbf{B} = \mu_o \mathbf{J} + \epsilon_o \mu_o \frac{\partial \mathbf{E}}{\partial t} . \tag{1.2.7}$$

Here $\epsilon_o (= 8.854 \times 10^{-12}$ F/m) and $\mu_o (= 4\pi \times 10^{-7}$ H/m) are the dielectric constant and the permeability of vacuum, $\rho(\mathbf{r}, t)$ and $\mathbf{J}(\mathbf{r}, t)$ are the charge density and the current density (field) which are obtained from the positions and the velocities of the individual charged particles of the plasma,

$$\rho(\mathbf{r}, t) = \sum_j q_j \delta[\mathbf{r} - \mathbf{r}_j(t)] \tag{1.2.8}$$

$$\mathbf{J}(\mathbf{r}, t) = \sum_j \mathbf{v}_j(t) q_j \delta[\mathbf{r} - \mathbf{r}_j(t)] , \tag{1.2.9}$$

where the summation is over all particles, each with charge q_j at position $\mathbf{r}_j(t)$ having velocity $\mathbf{v}_j(t)$. $\delta[\mathbf{r}-\mathbf{r}_j(t)]$ is the three-dimensional δ function given by $\delta[x-x_j(t)]\delta[y-y_j(t)]\,\delta[z-z_j(t)]$.

We note here that the particle velocity and position are functions of time (and the initial position and initial velocity), but not of the spatial coordinate $\mathbf{r}$, while fields are a function of time *and* the spatial coordinate (a definition of a field). The former is often called a Lagrangian variable while the latter (the field) an Eulerian variable.

We also note that the time derivative of Eq. (1.2.8) is the negative of the spatial derivative of Eq. (1.2.9); hence the charge density and the current density satisfy the continuity equation,

$$\frac{\partial \rho}{\partial t} + \nabla \cdot \mathbf{J} = 0 . \tag{1.2.10}$$

The source of the electric and magnetic fields are the charge and the current density as indicated by Eqs. (1.2.4) and (1.2.7). The electromagnetic field consists not only of the externally applied field but also of that internally produced by the charge and the current density carried by the individual plasma particles. The internally produced field consists of the field which is produced by the particle which is closest to the particular particle of immediate concern as well as the fields which are produced by the remainder of the particles. The influence of the field of the closest neighbor on the motion of the particle of immediate concern is called the Coulomb collision.

The canonical variables $\mathbf{P}, \mathbf{Q}$ and the Hamiltonian structure corresponding to the Lorentz equation of motion are

$$\mathbf{P}(t) = m\mathbf{v}(t) + q\mathbf{A}[\mathbf{r}(t), t] \tag{1.2.11}$$

$$\mathbf{Q}(t) = \mathbf{r}(t) \tag{1.2.12}$$

and the Hamiltonian H is given by,

$$H(\mathbf{P}, \mathbf{Q}, t) = \frac{1}{2m}(\mathbf{P} - q\mathbf{A})^2 + q\phi . \tag{1.2.13}$$

Here $\mathbf{A}(\mathbf{r}, t)$ is the magnetic vector potential,

$$\nabla \times \mathbf{A} = \mathbf{B} \tag{1.2.14}$$

and $\phi(\mathbf{r}, t)$ is the electric potential

$$\nabla \phi = -\mathbf{E} - \frac{\partial \mathbf{A}}{\partial t} . \tag{1.2.15}$$

It is straightforward to see that H satisfies

$$\frac{\partial H}{\partial P_j} = \dot{Q}_j , \tag{1.2.16}$$

where the subscript j designates jth component of the vector. To prove the other canonical equation,

$$\frac{\partial H}{\partial Q_j}(\equiv \nabla H) = -\dot{P}_j \tag{1.2.17}$$

one uses the vector identity,

$$\begin{aligned} &\nabla(\mathbf{P}-q\mathbf{A})^2 \\ &= \nabla[(\mathbf{P}-q\mathbf{A})\cdot(\mathbf{P}-q\mathbf{A})] \\ &= 2(\mathbf{P}-q\mathbf{A})\times[\nabla\times(\mathbf{P}-q\mathbf{A})] \\ &+ 2[(\mathbf{P}-q\mathbf{A})\cdot\nabla](\mathbf{P}-q\mathbf{A}), \end{aligned}$$

with

$$\dot{\mathbf{A}} = \frac{\partial \mathbf{A}}{\partial t} + (\mathbf{v}\cdot\nabla)\mathbf{A} \tag{1.2.18}$$

as well as the gauge relation

$$\mathbf{E} = -\frac{\partial \mathbf{A}}{\partial t} - \nabla\phi \tag{1.2.19}$$

and compares the result with the Lorentz equation of motion.

We note that Hamilton's equation of motion written in the Canonical form of Eqs. (1.2.16) and (1.2.17) applies also to the case of a relativistic particle.

1.3 The Vlasov Equation

The motion of an individual particle is governed by the electromagnetic field at the position of the particle. The position and the velocity of individual particles, which can be obtained by integrating the equation of motion, determine the charge density and current density as expressed in Eq. (1.2.8) and (1.2.9). The charge and current density in turn decide the new electromagnetic field through the Maxwell equations. Solving these coupled equations for the entire ensemble of plasma particles is an impossible task. To resolve this difficulty it is often convenient to introduce a statistical method based on the particle distribution function in "phase space".

If we note that the position of an individual particle depends on the initial velocity of the particle, we expect that the velocity of particles at a fixed position will vary even under the same force field.

In other words, even if we fix the position, there can be different velocities. This indicates that to fully describe a particle density, one must specify the six-dimensional phase space position, x, y, z, v_x, v_y, v_z. For N number of particles, one can define the six dimensional phase space density N (**r**, **v**, t)

$$N(\mathbf{r},\mathbf{v},t)=\sum_{j}^{N}\delta[\mathbf{r}-\mathbf{r}_j(t)]\,\delta[\mathbf{v}-\mathbf{v}_j(t)] \qquad (1.3.1)$$

in a way analogous to the definition of the charge density, Eq. (1.2.8), where the distribution function defined as Eq. (1.3.1) is called the Klimontovich distribution function. Naturally, for a relativistic particle, the momentum should be used instead of **v**. The number density $\rho(\mathbf{r},t)$ is obtained by integrating N over three dimensional velocity space,

$$n(\mathbf{r},t)=\int N(\mathbf{r},\mathbf{v},t)d\mathbf{v} \qquad (1.3.2)$$

$$=\sum_{j}^{N}\delta[\mathbf{r}-\mathbf{r}_j(t)]\;.$$

Integrating the right hand side further over the three-dimensional coordinate space, one obtains the total number of particles N.

One can derive the conservation law of N in phase space by taking the partial time derivative of N,

$$\frac{\partial N}{\partial t}=-\sum_{j}^{N}\dot{\mathbf{r}}_j\cdot\delta'[\mathbf{r}-\mathbf{r}_j(t)]\,\delta[\mathbf{v}-\mathbf{v}_j(t)]$$

$$-\sum_{j}^{N}\dot{\mathbf{v}}_j\cdot\delta[\mathbf{r}-\mathbf{r}_j(t)]\,\delta'[\mathbf{v}-\mathbf{v}_j(t)]$$

$$=-\sum_{j}^{N}\dot{\mathbf{r}}_j\cdot\frac{\partial}{d\mathbf{r}}\delta[\mathbf{r}-\mathbf{r}_j(t)]\,\delta[\mathbf{v}-\mathbf{v}_j(t)]$$

$$-\sum_{j}^{N}\dot{\mathbf{v}}_j\cdot\frac{\partial}{\partial\mathbf{v}}\delta[\mathbf{r}-\mathbf{v}_j(t)]\delta[\mathbf{v}-\mathbf{v}_j(t)]$$

$$=-\mathbf{v}\cdot\frac{\partial N}{\partial\mathbf{r}}-\dot{\mathbf{v}}\cdot\frac{\partial N}{\partial\mathbf{v}}$$

because $\dot{\mathbf{r}}_j=\mathbf{v}_j=\mathbf{v}$ as well as $\dot{\mathbf{v}}_j=\dot{\mathbf{v}}$ from the property of the δ function. Hence the conservation law for N in phase space reads,

$$\frac{\partial N}{\partial t}+\mathbf{v}\cdot\frac{\partial N}{\partial\mathbf{r}}+\dot{\mathbf{v}}\cdot\frac{\partial N}{\partial\mathbf{v}}=0\;. \qquad (1.3.3)$$

Equation (1.3.3) is called the Klimontovich equation. The fine-grained distribution function of Klimontovich is precise in describing the microscopic states of the many-particle system. However, it would not by itself correspond to the coarse-grained distribution function we observe in the macroscopic world. Hence, we introduce a coarse-grained single-particle distribution function $f(\mathbf{r}, \mathbf{v}, t)$ defined as the ensemble average over the j th particle,

$$f(\mathbf{r}, \mathbf{v}, t) = \langle N(\mathbf{r}, \mathbf{v}, t) \rangle_j \qquad (1.3.4)$$

If we take ensemble average of Eq. (1.3.3) and ignore correlations between two (and more) particles, we have

$$\frac{\partial f}{\partial t} + \mathbf{v} \cdot \frac{\partial f}{\partial \mathbf{r}} + \dot{\mathbf{v}} \cdot \frac{\partial f}{\partial \mathbf{v}} = 0 . \qquad (1.3.5)$$

Although Eq. (1.3.5) formally resembles Eq. (1.3.3), it has a different physical content. In particular the solution of Eq. (1.3.5) does not involve the positions and velocities of the jth particles. Here, in view of Eqs. (1.3.1) and (1.3.2), we note that

$$\int f(\mathbf{r}, \mathbf{v}, t) d\mathbf{v} = n(\mathbf{r}, t) \qquad (1.3.6)$$

is the number density in the coordinate space. The first moment gives the particle flux density,

$$n(\mathbf{r}, t)\mathbf{v}(\mathbf{r}, t) = \int \mathbf{v} f(\mathbf{r}, \mathbf{v}, t) d\mathbf{v} . \qquad (1.3.7)$$

Here $\mathbf{v}(\mathbf{r}, t)$ is the velocity field which should be distinguished from the phase space variable $\mathbf{v}$ and $\int d\mathbf{v}$ is the volume integral in $\mathbf{v}$ space. If one uses the canonical variables $\mathbf{P}$, $\mathbf{Q}$ as the phase space variables, using Eqs. (1.2.16) and (1.2.17), Eq. (1.3.5) may be expressed as

$$\frac{\partial f}{\partial t} + \{f, H\} = 0 , \qquad (1.3.8)$$

where $\{f, H\}$ is called the Poisson bracket defined as

$$\{f, H\} = \frac{\partial H}{\partial P_j} \frac{\partial f}{\partial Q_j} - \frac{\partial H}{\partial Q_j} \frac{\partial f}{\partial P_j} . \qquad (1.3.9)$$

Equation (1.3.8) shows that if $f = f(H)$, $\{f, H\} = 0$, hence $f(H)$ is a stationary solution of Eq. (1.3.5).

If one uses the Hamiltonian in the electromagnetic field, Eq. (1.2.13), Eq. (1.3.8) gives,

$$\frac{\partial f}{\partial t} + \mathbf{v} \cdot \frac{\partial f}{\partial \mathbf{r}} + \frac{q}{m}(\mathbf{E} + \mathbf{v} \times \mathbf{B}) \cdot \frac{\partial f}{\partial \mathbf{v}} = 0 . \qquad (1.3.10)$$

Equation (1.3.10) is called the Vlasov equation or the collisionless Boltzmann equation.

1.4 Characteristic Scales in a Plasma

Cyclotron Frequency, ω_c

A charged particle with a velocity in the direction perpendicular to a static and uniform magnetic field $\mathbf{B}_0$ executes a circular motion with a fixed frequency. If we take a time derivative of the Lorentz equation of motion, (1.2.1) without an electric field,

$$m\ddot{\mathbf{v}} = q\dot{\mathbf{v}} \times B_0$$

and substitute this back into Eq. (1.2.1), we have

$$m\ddot{\mathbf{v}} = \frac{q^2}{m}(\mathbf{v} \times \mathbf{B}_0) \times \mathbf{B}_0 \tag{1.4.1}$$

$$= -\frac{q^2}{m} B_0^2 \mathbf{v} \ .$$

Equation (1.4.1) indicates a periodic motion in time with an angular frequency given by qB_0/m. This frequency is called the cyclotron frequency, usually written as ω_c. For electrons,

$$\omega_{ce} = 1.76 \times 10^{11} B_0(T)\ s^{-1} \ . \tag{1.4.2}$$

Writing as x and y components in the direction perpendicular to $\mathbf{B}_0$,

$$\dot{v}_x = -\frac{e}{m} v_y B_0 \ .$$

Hence, if $v_x = v_\perp \cos\omega_{ce}t$, $v_y = v_\perp \sin\omega_{ce}t$. This is a circular motion with the right hand direction pointing along the z axis, namely the direction of the magnetic field $\mathbf{B}_0$. In a similar manner, the ions rotate to the left. In either case, the current due to the motion is diamagnetic, (see Fig. 1.1).

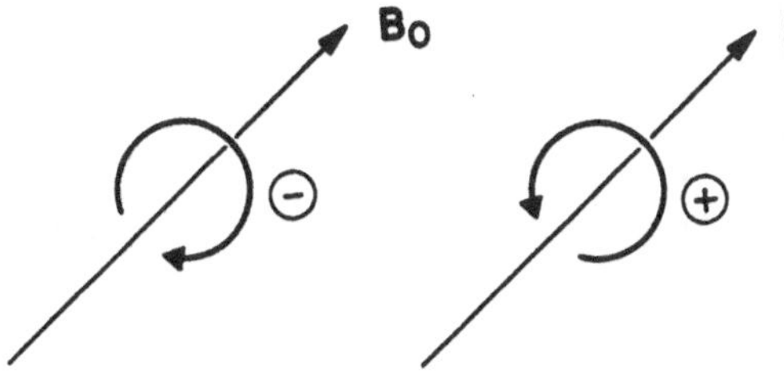

Fig. 1.1. Cyclotron motion of charged particles

Plasma Frequency, ω_p

A plasma maintains a charge neutrality owing to the free electrons. When electrons are displaced from ions by a small distance, they are pulled back by the Coulomb force of the ions and exhibit an oscillation due to their inertia. This oscillation is called the plasma oscillation. Unlike cyclotron oscillation, the plasma oscillation is a "collective" oscillation indicating the presence of dispersion.

Let us consider a square bulk of plasma and displace the electrons perpendicular to one of the planes (see Fig. 1.2) by a distance Δ. This will induce electric charge density per unit area given by $en_o\Delta$. Hence, from Poisson's equation, the electric field E generated by the displacement is given by

$$E = \frac{en_o}{\epsilon_o}\Delta, \tag{1.4.3}$$

where n_o is the number density of electrons, e is the electron charge and $\epsilon_o(=8.854\times10^{-12}\mathrm{F/m})$ is the dielectric constant of the vacuum. The equation of motion for an electron in the bulk of the displaced electrons, is then given by

$$m_e\ddot{\Delta} = -\frac{e^2n_o}{\epsilon_o}\Delta \tag{1.4.4}$$

where m_e is the electron mass. This equation indicates a periodic motion of the displacement Δ with a frequency given by $(e^2n_o/\epsilon_om_e)^{1/2}$. This is called the plasma frequency, ω_p

$$\omega_p = \left(\frac{e^2n_o}{\epsilon_om_e}\right)^{1/2} = 56\sqrt{n_o(m^{-3})}\ \ s^{-1}. \tag{1.4.5}$$

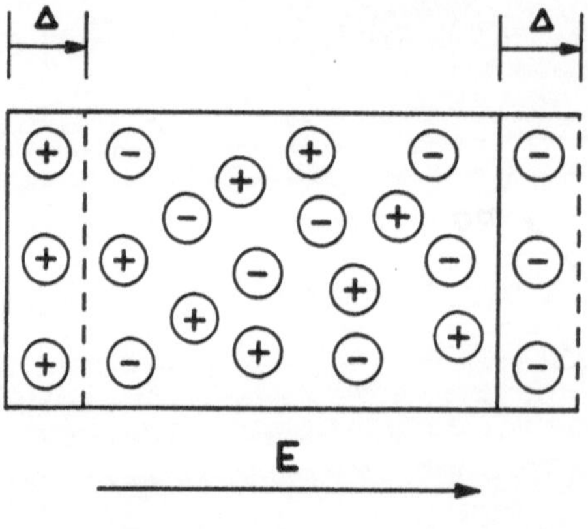

Fig. 1.2. Oscillation of displaced electron charge Δ

Thermal Speed, v_T

When the velocity distribution function is given by a Maxwellian with a temperature T (in Joule units),

$$f_o(\mathbf{v}) = (\frac{m}{2\pi T})^{3/2} \exp\left(-\frac{mv^2}{2T}\right), \qquad (1.4.6)$$

the average of squared velocity in one component in Cartesian coordinates, say v_x^2, is given by

$$<v_x^2> = \int_{-\infty}^{\infty} v_x^2 f_o(\mathbf{v}) dv_x dv_y dv_z$$

$$= \frac{T}{m} \left(= <v_y^2> = <v_z^2> \right). \qquad (1.4.7)$$

The square root of $<v_x^2>$ is called the thermal speed v_T. For electrons,

$$v_{Te} = (T/m_e)^{1/2}$$

$$= 4.2 \times 10^5 \sqrt{T(eV)}\ m/s \qquad (1.4.8)$$

For the same temperature, the ion thermal speed is smaller than the electron thermal speed by the square root of the mass ratio.

Larmor Radius, ρ

As we saw in the derivation of the cyclotron frequency, a charged particle with a velocity $v_\perp$ perpendicular to the magnetic field exhibits a circular motion. The radius of rotation is given by $v_\perp/\omega_c$. This quantity is called the Larmor radius, ρ,

$$\rho = \frac{v_\perp}{\omega_c}. \qquad (1.4.9)$$

For the same energy, the ion Larmor radius is larger than the electron Larmor radius by the square root of the mass ratio.

Debye Shielding Length, λ_D

Strictly speaking, plasma oscillation appears only for a cold plasma. When the effect of a finite temperature T_e is included, in response to the electric field, the electron gas tends to redistribute itself to form the Boltzmann distribution,

$$n = n_o \exp(e\phi/T_e).$$

$$\simeq n_o(1+e\phi/T_e) \quad \text{for} \quad e\phi/T_e \ll 1 , \tag{1.4.10}$$

where ϕ is the electric potential. Poisson's equation then reads,

$$\frac{d^2\phi}{dx^2} = -\frac{e}{\epsilon_o}(n_i - n_e)$$

$$\simeq \frac{e}{\epsilon_o}\frac{e\phi}{T_e}n_o .$$

Solving this for ϕ, we have

$$\phi = \exp(-k_D x) , \tag{1.4.11}$$

where

$$k_D^2 = \frac{e^2 n_o}{\epsilon_o T_e} = \frac{\omega_p^2}{v_{Te}^2} . \tag{1.4.12}$$

Equation (1.4.11) shows that the electric field, instead of oscillating at ω_p, becomes static and is shielded beyond a distance of $1/k_D$. k_D is called the Debye wavenumber and $\lambda_D = 2\pi/k_D$ is called the Debye (wave) length. This implies that the plasma oscillation disappears (or is shielded) if the wavelength of the oscillation is shorter the λ_D.

Collision Frequency ν and Mean Free Path λ_m

When two particles in a plasma approach within a distance r_o of each other, such that the kinetic energy in the relative motion of each

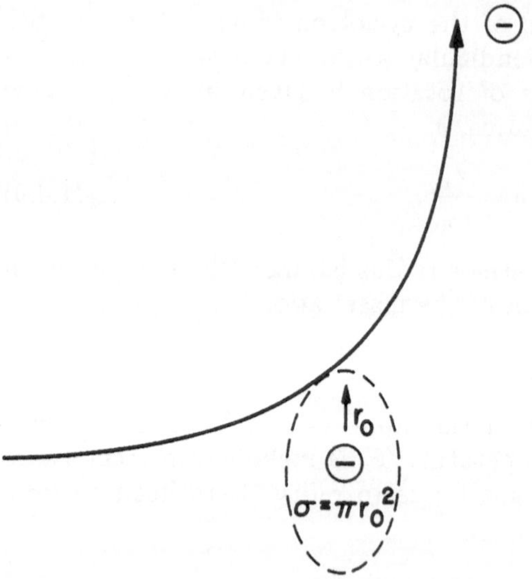

Fig. 1.3. Collision between two charged particles

particle becomes comparable to the Coulomb potential energy, the particle trajectories are deflected (see Fig. 1.3). This process is called a Coulomb collision. If we take into account Coulomb collisions, the Vlasov equation [Eq. (1.3.10)] is modified because it assumes only the averaged electromagnetic field. The Coulomb collision frequency ν is given by $\nu = \sigma n v$ where σ is the collision cross-section, n the number density of the particle and v is the relative speed between the colliding particles. Since the Coulomb field falls off slowly as r^{-1}, the cross-section of a charged particle diverges. However, if we take into account the Debye shielding effect due to the background plasma, the cross-section converges and the electron-ion collision frequency ν_{ei} is given by

$$\nu_{ei} = \frac{2\pi\sqrt{2\pi}}{3}\frac{\omega_{pe}}{n\lambda_D^3}\ell n\Lambda$$

$$= 2.9\times 10^{-12}\frac{n(m^{-3})\ell n\Lambda}{T_e(eV)^{3/2}}\ s^{-1}, \tag{1.4.13}$$

where $\Lambda(= \frac{12\pi(\epsilon_o T/e^2)^{3/2}}{n_e^{1/2}})$ is nine times the number of particles in the Debye sphere and $\ell n\Lambda$, which is called the Coulomb logarithm, has a value typically of 10 ~ 20. For example, the values of λ_D in the ionosphere, plasmasphere, and in the earth's magnetosphere have typical values of $\sim 5\times 10^{-2}$, 0.5 and 300 m respectively, while the number densities are $\sim 10^{11}$, 10^9, and $10^7\ m^{-3}$ respectively. The corresponding plasma parameter, $(n\lambda_D^3)^{-1}$, for these cases is $\sim 10^{-7}$, 10^{-8} and 10^{-14}. The plasma frequency, ω_{pe}, in these locations is typically 10^7, 10^6 and $10^5 s^{-1}$. Hence the Coulomb collision frequency is much smaller than most other characteristic frequencies in space plasmas.

The electron-electron collision frequency is $2\sqrt{2}$ times ν_{ei}, i.e., $\nu_{ee} = 2\sqrt{2}\,\nu_{ei}$ while the ion-ion collision frequency ν_{ii} is $(m_e/m_i)^{1/2}$ times ν_{ee}, i.e., $\nu_{ii} = (m_e/m_i)^{1/2}\ \nu_{ee}$ and the ion-electron collision frequency ν_{ie} is approximately given by $\nu_{ie} \simeq (m_e/m_i)\ \nu_{ei}$, since an ion is more difficult to deflect. The ratio of the thermal velocity $v_T = \sqrt{T/m}$ to the collision frequency ν is called the mean free path λ_m,

$$\lambda_m = \frac{v_T}{\nu} = \sqrt{\frac{T}{m}}\frac{1}{\nu}, \tag{1.4.14}$$

where T is the temperature in energy units and m is the mass of the species. The mean free path has different values depending on the species.

1.5 Conservation Laws

If we take the divergence of Eq. (1.2.7) and substitute Eq. (1.2.4), we obtain the equation of continuity,

$$\frac{\partial \rho}{\partial t} + \nabla \cdot \mathbf{J} = 0 . \tag{1.5.1}$$

We note that this equation can also be derived by taking the partial time derivative of Eq. (1.2.8) and by substituting the result into Eq. (1.2.9) as in Eq. (1.2.10). If we integrate this equation over a geometric volume V,

$$\frac{\partial}{\partial t} \int \rho dV + \oint \mathbf{J} \cdot d\mathbf{S} = 0 . \tag{1.5.2}$$

The quantity $\int \rho dV (= N)$ is the total number of charged particles within the volume V while $\oint \mathbf{J} \cdot d\mathbf{S}$ is the sum of the current (or flow of the charge) which crosses the surface of the volume V. Equation (1.5.2) expresses the fact that the time rate change of the total charge in the volume V is given by the balance of the charge flow across the entire surface. Hence Eq. (1.5.1) is also called the equation of charge conservation.

In general, a conservation law for a density a can be written in terms of its flow vector $\mathbf{A}$ as

$$\frac{\partial a}{\partial t} + \nabla \cdot \mathbf{A} = 0. \tag{1.5.3}$$

We note that if the quantity a is a vector, A becomes a tensor, and so forth.

Conservation of Particle Density

The equation of particle conservation can be obtained by integrating the Vlasov equation (1.3.10) over the entire velocity space. We note from Eqs. (1.3.6) and (1.3.7)

$$\int \frac{\partial f}{\partial t} d\mathbf{v} = \frac{\partial}{\partial t} \int f d\mathbf{v} = \frac{\partial}{\partial t} n(\mathbf{r}, t)$$

$$\int \mathbf{v} \cdot \frac{\partial f}{\partial \mathbf{r}} d\mathbf{v} = \nabla \cdot \int \mathbf{v} f d\mathbf{v}$$

$$= \nabla \cdot [n(\mathbf{r}, t) \mathbf{v}(\mathbf{r}, t)]$$

while

$$\int (\mathbf{E} + \mathbf{v} \times \mathbf{B}) \cdot \frac{\partial f}{\partial \mathbf{v}} d\mathbf{v} = 0.$$

Thus the particle (mass) conservation equation reads,

$$\frac{\partial}{\partial t}n(\mathbf{r},t)+\nabla\cdot[n(\mathbf{r},t)\mathbf{v}(\mathbf{r},t)]=0\,. \tag{1.5.4}$$

Conservation of Energy Density

The equation of energy conservation is obtained by multiplying the Vlasov equation by $\frac{1}{2}mv^2$ and integrating the result over the velocity space. In order to perform the integration, it is convenient to introduce an independent variable $\delta\mathbf{v}$ such that

$$\mathbf{v}=\mathbf{v}(\mathbf{r},t)+\delta\mathbf{v}\,, \tag{1.5.5}$$

where $\mathbf{v}(\mathbf{r},t)$ is the velocity field as defined in Eq. (1.3.7) and $\delta\mathbf{v}$ is the deviation from the average velocity $\mathbf{v}(\mathbf{r},t)$. By definition

$$\int \delta\mathbf{v}f(\mathbf{v},\mathbf{r},t)d\mathbf{v}=0\,. \tag{1.5.6}$$

Then

$$\frac{1}{2}m\frac{\partial}{\partial t}\int v^2 f d\mathbf{v}$$

$$=\frac{\partial}{\partial t}\left[\frac{1}{2}mn(r,t)v(\mathbf{r},t)^2+\frac{1}{2}mn(\mathbf{r},t)<\delta v^2>\right]\equiv\frac{\partial}{\partial t}E_K\,, \tag{1.5.7}$$

where $<>$ indicates the average and E_K is the kinetic energy density, while

$$\frac{1}{2}m\frac{\partial}{\partial x_i}\int v_i v^2 f d\mathbf{v}$$

$$=\frac{\partial}{\partial x_i}\left[\frac{1}{2}mn(\mathbf{r},t)v(\mathbf{r},t)^2 v_i(\mathbf{r},t)\right.$$

$$+\frac{1}{2}mn(\mathbf{r},t)\Big(<\delta v^2>+<\delta v_i\delta v_j>\Big)v_i(\mathbf{r},t)$$

$$\left.+\frac{1}{2}mn(\mathbf{r},t)<\delta v^2\delta v_i>\right]$$

$$\equiv(\nabla\cdot\mathbf{P}_K)_i\,, \tag{1.5.8}$$

where $\mathbf{P}_K$ is the kinetic energy flow vector. In the expression for $\mathbf{P}_K$, the first term represents the flow of macroscopic kinetic energy. The second term represents the flow of the internal energy carried by the velocity field, the convection. These two terms vanish in the frame moving with the velocity field $\mathbf{v}(\mathbf{r},t)$. The last term, which need not vanish even in such a frame, represents the energy carried by the (heat) conduction. The last term of the Vlasov equation gives

$$\int \frac{1}{2}mv^2 \frac{q}{m}(\mathbf{E}+\mathbf{v}\times\mathbf{B})\cdot\frac{\partial f}{\partial \mathbf{v}}d\mathbf{v} = -qn(\mathbf{r},t)\mathbf{v}(\mathbf{r},t)\cdot\mathbf{E}. \quad (1.5.9)$$

Taking into account both electrons and ions we have

$$\frac{\partial}{\partial t}\varepsilon_K + \nabla\cdot\mathbf{P}_K = \mathbf{J}\cdot\mathbf{E}, \quad (1.5.10)$$

where

$$\mathbf{J}(\mathbf{r},t) = en(\mathbf{r},t)[\mathbf{v}_i(\mathbf{r},t) - \mathbf{v}_e(\mathbf{r},t)] \quad (1.5.11)$$

is the electric current density, $\mathbf{v}_i$, $\mathbf{v}_e$ and e are the velocity fields of ions and electrons and the magnitude of electronic charge, respectively. We note that, unlike the mass conservation, the energy conservation equation does not close itself within the kinetic energy, but couples with the electromagnetic energy through $\mathbf{J}\cdot\mathbf{E}$. $\mathbf{J}\cdot\mathbf{E}$ can be written in the form of a conservation equation (Poynting Theorem) using Maxwell's equations by constructing $\mathbf{E}\cdot$Eq. (1.2.7) $-$ $\mathbf{B}\cdot$Eq. (1.2.5) and by using the vector formula

$$\nabla\cdot(\mathbf{E}\times\mathbf{B}) = \mathbf{B}\cdot\nabla\times\mathbf{E} - \mathbf{E}\cdot\nabla\times\mathbf{B},$$

$$\mathbf{J}\cdot\mathbf{E} = -\frac{\partial}{\partial t}\left(\frac{1}{2}\epsilon_0 E^2 + \frac{1}{2\mu_0}B^2\right) - \nabla\cdot\left(\frac{1}{\mu_0}\mathbf{E}\times\mathbf{B}\right) \equiv -\left[\frac{\partial}{\partial t}\varepsilon_F + \nabla\cdot\mathbf{P}_F\right],$$

where ε_F and $\mathbf{P}_F$ are the energy density and the power flow of the electromagnetic field. Combining this with Eq. (1.5.10) gives the equation of energy density conservation,

$$\frac{\partial}{\partial t}(\varepsilon_F + \varepsilon_K) + \nabla\cdot(\mathbf{P}_F + \mathbf{P}_K) = 0. \quad (1.5.12)$$

Conservation of Momentum Density

Since momentum is a vector, let us take one (i) component and derive the conservation equation,

$$m\frac{\partial}{\partial t}\int v_i f d\mathbf{v} = \frac{\partial}{\partial t}[mn(\mathbf{r},t)v_i(\mathbf{r},t)] \equiv \frac{\partial}{\partial t}\pi_{Ki} \tag{1.5.13}$$

$$m\frac{\partial}{dx_j}\int v_i v_j f d\mathbf{v} = \frac{\partial}{\partial x_j}\Big[mn(\mathbf{r},t)v_i(\mathbf{r},t)v_j(\mathbf{r},t)$$

$$+ mn(\mathbf{r},t)<\delta v_i \delta v_j>\Big] \equiv \frac{\partial}{\partial x_j}\Gamma_{Kij} \,. \tag{1.5.14}$$

Here mn $<\delta v_i \delta v_j> \equiv p_{ij}$ is the pressure tensor. If the distribution of $\delta\mathbf{v}$ is Maxwellian with the temperature T, Γ_{ij} becomes a scalar and is given by

$$\Gamma_{ij} = nT\overleftrightarrow{\mathbf{I}}, \tag{1.5.15}$$

where $\overleftrightarrow{\mathbf{I}}$ is the unit tensor,

$$\overleftrightarrow{\mathbf{I}} = \begin{pmatrix} 1 & 0 & 0 \\ 0 & 1 & 0 \\ 0 & 0 & 1 \end{pmatrix}.$$

Finally

$$\int mv_i \frac{q}{m}(\mathbf{E}+\mathbf{v}\times\mathbf{B})\cdot\frac{\partial f}{\partial \mathbf{v}} d\mathbf{v}$$

$$= -q\Big[n(\mathbf{r},t)\mathbf{E} + n(\mathbf{r},t)\mathbf{v}(\mathbf{r},t)\times\mathbf{B}\Big]_i \,. \tag{1.5.16}$$

If we sum (1.5.16) over electrons and ions, we obtain

$$\sum_{i,e}\int m\mathbf{v}\frac{q}{m}(\mathbf{E}+\mathbf{v}\times\mathbf{B})\cdot\frac{\partial f}{\partial \mathbf{v}} d\mathbf{v}$$

$$= -\rho_a\mathbf{E} - \mathbf{J}\times\mathbf{B}, \tag{1.5.17}$$

where $\rho_a \Big[= e[n_i(\mathbf{r},t) - n_e(\mathbf{r},t)]\Big]$ is the ambipolar charge.

Like the energy conservation equation, the momentum conservation equation does not close itself within the kinetic part only. This indicates that the electromagnetic field also carries momentum. To obtain the total momentum conservation equation, one needs to express $\rho_a\mathbf{E}$ and $\mathbf{J}\times\mathbf{B}$ in the form of a conservation equation. From Eq. (1.2.4), we note

$$\rho_a\mathbf{E} = \epsilon_o(\nabla\cdot\mathbf{E})\mathbf{E} = \epsilon_o\{\nabla\cdot(\mathbf{EE}) - (\mathbf{E}\cdot\nabla)\mathbf{E}\}\,, \tag{1.5.18}$$

where use is made of the identity for the divergence of a tensor **AB**,

$$\nabla\cdot(\mathbf{AB}) = \mathbf{A}(\nabla\cdot\mathbf{B}) + (\mathbf{B}\cdot\nabla)\mathbf{A}\,.$$

while

$$(\mathbf{E}\cdot\nabla)\mathbf{E} = \nabla\frac{E^2}{2} - \mathbf{E}\times(\nabla\times\mathbf{E})$$

$$= \nabla\frac{E^2}{2} + \mathbf{E}\times\frac{\partial\mathbf{B}}{\partial t}. \qquad (1.5.19)$$

Thus

$$\rho_a\mathbf{E} = \epsilon_o\{\nabla\cdot(\mathbf{EE}) - \nabla\frac{E^2}{2} - \mathbf{E}\times\frac{\partial\mathbf{B}}{\partial t}\}. \qquad (1.5.20)$$

Similarly

$$\mathbf{J}\times\mathbf{B} = \frac{1}{\mu_o}(\nabla\times\mathbf{B})\times\mathbf{B} - \epsilon_o\frac{\partial\mathbf{E}}{\partial t}\times\mathbf{B}$$

$$= -\nabla\frac{B^2}{2\mu_o} + \frac{1}{\mu_o}(\mathbf{B}\cdot\nabla)\mathbf{B} - \epsilon_o\frac{\partial\mathbf{E}}{\partial t}\times\mathbf{B}$$

$$= -\nabla\frac{B^2}{2\mu_o} + \frac{1}{\mu_o}\nabla\cdot(\mathbf{BB}) - \epsilon_o\frac{\partial\mathbf{E}}{\partial t}\times\mathbf{B}. \qquad (1.5.21)$$

Combining Eqs. (1.5.20) and (1.5.21), we obtain the desired structure,

$$\rho_a\mathbf{E} + \mathbf{J}\times\mathbf{B} = -\epsilon_o\frac{\partial}{\partial t}(\mathbf{E}\times\mathbf{B})$$

$$-\nabla\cdot\left[\left(\frac{\epsilon_o E^2}{2} + \frac{B^2}{2\mu_o}\right)\overleftrightarrow{\mathbf{I}} - (\epsilon_o\mathbf{EE} + \frac{1}{\mu_o}\mathbf{BB})\right]$$

$$\equiv -\frac{\partial}{\partial t}\pi_F - \nabla\cdot\overleftrightarrow{\Gamma}_F. \qquad (1.5.22)$$

If we combine this result with Eqs. (1.5.14) and (1.5.17), we have the equation of momentum density conservation,

$$\frac{\partial}{\partial t}\left(\pi_K + \pi_F\right) + \nabla\cdot\left(\overleftrightarrow{\Gamma}_K + \overleftrightarrow{\Gamma}_F\right). \qquad (1.5.23)$$

1.6 Magnetohydrodynamic Equations

Derivation

In order to describe the behavior of a collisionless plasma it is ideal to use the Vlasov equation. However, the equation is difficult to use, in particular for an inhomogeneous plasma where the full six-dimensional effects become important. There have been various attempts to derive a simplified set of equations to resolve the difficulty. One such attempt treats the plasma as a fluid (rather than as a group of discrete charged particles). The simplest fluid equations (although still difficult to solve) are called the magnetohydrodynamic (or MHD) equations.

These equations are derived by assuming that the plasma retains a Maxwellian distribution. Hence, the equations are applicable for processes in which the temporal changes are slower than the slowest characteristic frequency of a single particle (i.e., the ion cyclotron frequency, ω_{ci}) and the spatial variations are smoother than the characteristic scale length of a single particle (i.e., the ion Larmor radius ρ_i). Ideally, the Maxwellian distribution is maintained by inter-particle collisions; hence, if the process under consideration occurs during a time period shorter than a collision time, the assumption of a Maxwellian distribution, and hence the MHD equations, become invalid. In most space plasmas, the inter-particle collision time is very long. For example, the mean free path of a solar wind particle is approximately 1 AU (the distance between the sun and the earth). Hence the natural question is the applicability of MHD to a space plasma.

There are two empirical ways to justify the applicability. One is that anomalous collisions produced by various microinstabilities, some of which will be introduced in the second volume, tend to keep the distribution close to Maxwellian. The other is to assume that the effects which arise from any deviation from the Maxwellian are small. The latter argument is equivalent to the neglect of thermal transport.

In order to allow the presence of a hydrodynamic flow, we assume a Maxwellian with an average velocity $<\mathbf{v}>$, which is nothing but the velocity field $\mathbf{v}(\mathbf{r}, t)$ introduced in Eq. (1.3.7),

$$f(\mathbf{r}, \mathbf{v}, t) = n(\mathbf{r}, t)\left(\frac{m}{2\pi T}\right)^{3/2} \exp\left[-\frac{m(\mathbf{v} - <\mathbf{v}>)^2}{2T}\right]. \qquad (1.6.1)$$

We substitute this distribution function to the zeroth, first and the second moment equations derived in Section 1.5.

First, we recognize the zeroth moment equation as being the continuity equation (1.5.4). When the electron and ion mass is multiplied by the electron and ion continuity equations and summed together, we have,

$$\frac{\partial n}{\partial t} + \nabla \cdot (n\mathbf{v}) = 0, \qquad (1.6.2)$$

where $\mathbf{v}=\dfrac{m_i<\mathbf{v}_i>+m_e<\mathbf{v}_e>}{m_i+m_e}$ is the velocity field of the center of gravity.

In the first moment equation, we note that the Maxwellian distribution assumed here gives an isotropic pressure $p=nT$. Hence the *i*th component of the moment equation reads,

$$\frac{\partial}{\partial t}(mn<v_i>)+\frac{\partial}{\partial x_j}(mn<v_i><v_j>)+\frac{\partial p}{\partial x_i}$$

$$=qnE_i+qn(\mathbf{v}\times\mathbf{B})_i\,. \tag{1.6.3}$$

Now

$$\frac{\partial}{\partial t}(mn<v_i>)=m<v_i>\frac{\partial n}{\partial t}+mn\frac{\partial<v_i>}{\partial t}$$

and

$$\frac{\partial}{\partial x_j}(mn<v_i><v_j>)=m<v_i>\nabla\cdot(n<\mathbf{v}>)+mn<\mathbf{v}>\cdot\nabla<v_i>\,.$$

Using the continuity equation (1.6.2), and noting the "convective derivative",

$$\frac{d}{dt}=\frac{\partial}{\partial t}+\mathbf{v}\cdot\nabla\,, \tag{1.6.4}$$

we have from Eq. (1.6.3),

$$mn\frac{d\mathbf{v}}{dt}=-\nabla p+qn(\mathbf{E}+\mathbf{v}\times\mathbf{B})\,. \tag{1.6.5}$$

We now use the Maxwellian distribution into the second moment equation (1.5.8). We note that $<\delta v^2>=<\delta v_x^2>+<\delta v_y^2>+<\delta v_z^2>=3T$, while, $<\delta v_i\delta v_j>=\delta_{ij}T$, $<\delta v^2\delta v_i>=0$, where $\delta_{ij}=1$ (if i=j), $=0$ (if $i\neq j$). Then the second moment equation reduces to

$$mn\,v_i\frac{\partial v_i}{\partial t}+\frac{1}{2}mv^2\frac{\partial n}{\partial t}+\frac{3}{2}\frac{\partial p}{\partial t}$$

$$+mn\,v_i(v_j\frac{\partial v_i}{\partial x_j})+\frac{1}{2}mv^2\frac{\partial}{\partial x_j}(nv_j)$$

$$+\frac{5}{2}v_i\frac{\partial p}{\partial x_i}+\frac{5}{2}p\frac{\partial v_i}{\partial x_i}=\mathbf{J}\cdot\mathbf{E}\,.$$

We note that the second and the fifth terms cancel by virtue of the continuity equation. Similarly in view of the equation of motion, (1.6.5), we note that the first, fourth, 2/2 part of the sixth are equal to

$\rho\mathbf{v}\cdot\mathbf{E}+\mathbf{v}\cdot(\mathbf{J}\times\mathbf{B})$. Now for each charged species, $\rho\mathbf{v}=\mathbf{J}$, hence $\mathbf{v}\cdot\mathbf{J}\times\mathbf{B}=0$. Hence, these terms cancel with $\mathbf{J}\cdot\mathbf{E}$ on the right hand side. The resultant equations give, together with Eq. (1.6.4),

$$\frac{3}{2}\frac{dp}{dt}=\frac{5}{2}T\frac{dn}{dt}.$$

This relation holds independently for both the electron pressure p_e and ion pressure p_i. If we define the total pressure $p=p_i+p_e$ and sum over species,

$$\frac{3}{2}\frac{dp}{dt}=\frac{5}{2}(T_i+T_e)\frac{dn}{dt},$$

where T_i and T_e are the ion and electron temperatures. Dividing both sides by T_e+T_i and using $p=(T_i+T_e)n$, we have the relation between the number density and the total plasma pressure,

$$3\frac{d}{dt}\ell n p=5\frac{d}{dt}\ell n n. \tag{1.6.6}$$

This relation indicates the adiabatic law for an ideal gas with the adiabatic constant $\gamma=5/3$. The appearance of the adiabatic relation (1.6.6) is the natural consequence of the assumption of the Maxwellian distribution because the absence of the third moment automatically inhibits the heat conduction.

So far we did not specify the species of the equations where the moments were taken. We now derive the equation of motion for the plasma treated as one fluid. For this purpose we write Eq. (1.6.5) for ions and electrons with subscripts i and e,

$$m_i\frac{d\mathbf{v}_i}{dt}=e(\mathbf{E}+\mathbf{v}_i\times\mathbf{B})-\frac{\nabla p_i}{n} \tag{1.6.7}$$

$$m_e\frac{d\mathbf{v}_e}{dt}=-e(\mathbf{E}+\mathbf{v}_e\times\mathbf{B})-\frac{\nabla p_e}{n}. \tag{1.6.8}$$

We introduce again the one fluid velocity field $\mathbf{v}(\mathbf{r},t)$ defined as the velocity for the center of gravity,

$$\mathbf{v}(\mathbf{r},t)=\frac{m_i\mathbf{v}_i+m_e\mathbf{v}_e}{m_i+m_e}$$

$$\simeq\mathbf{v}_i+m_e\frac{\mathbf{v}_e}{m_i}. \tag{1.6.9}$$

The equation of motion for this fluid with velocity $\mathbf{v}$, and mass density $\rho_m=(m_i+m_e)n\simeq m_i n$ is given by adding Eq. (1.6.7) and (1.6.8),

$$m_i n\frac{d\mathbf{v}}{dt}=\mathbf{J}\times\mathbf{B}-\nabla p, \tag{1.6.10}$$

where $p(=p_i+p_e)$ is the total pressure. Equation (1.6.10) indicates that the acceleration of the center of gravity of the plasma as one fluid is given by $\mathbf{J}\times\mathbf{B}$ and ∇p and that the electric field is not responsible for the acceleration. This is because a plasma does not have an average electric charge. The role of the electric field can be found by constructing the difference between Eqs. (1.6.7) and (1.6.8). To construct $\mathbf{v}$ from $\mathbf{v}_i$ and $\mathbf{v}_e$ in the $\mathbf{v}\times\mathbf{B}$ term, we multiply Eq. (1.6.8) by m_e/m_i and subtract from it Eq. (1.6.7) multiplied by e^{-1},

$$\frac{m_i}{e}\frac{d\mathbf{v}_i}{dt}-\frac{m_e^2}{em_i}\frac{d\mathbf{v}_e}{dt}=\mathbf{E}+\mathbf{v}\times\mathbf{B}+\frac{\nabla p_e}{en}\frac{m_e}{m_i}-\frac{\nabla p_i}{en}. \tag{1.6.11}$$

From Eq. (1.6.10), we can write the ion inertia term as

$$m_i n\frac{d\mathbf{v}_i}{dt}=\mathbf{J}\times\mathbf{B}-\nabla p_i-\nabla p_e-m_e n\frac{d\mathbf{v}_e}{dt}.$$

Dividing both sides by *en*,

$$\frac{m_i}{e}\frac{d\mathbf{v}_i}{dt}=\frac{\mathbf{J}\times\mathbf{B}}{en}-\frac{\nabla p_i}{en}-\frac{\nabla p_e}{en}-\frac{m_e}{e}\frac{d\mathbf{v}_e}{dt}. \tag{1.6.12}$$

Substituting Eq. (1.6.12) into (1.6.11), and ignoring the term with m_e/m_i, we have

$$\mathbf{E}+\mathbf{v}\times\mathbf{B}=-\frac{m_e}{e}\frac{d\mathbf{v}_e}{dt}-\frac{\nabla p_e}{en}+\frac{\mathbf{J}\times\mathbf{B}}{en}. \tag{1.6.13}$$

Furthermore, if we had kept the electron-ion collision effect, the loss of electron momentum would have contributed to the resistivity $\eta(=\frac{\nu_e m_e}{e^2 n})$ [see Eq. (1.12.1)]. Then Eq. (1.6.13) would be modified to

$$\mathbf{E}+\mathbf{v}\times\mathbf{B}=-\frac{m_e}{e}\frac{d\mathbf{v}_e}{dt}-\frac{\nabla p_e}{en}+\frac{\mathbf{J}\times\mathbf{B}}{en}+\eta\mathbf{J}. \tag{1.6.14}$$

This equation is called the generalized Ohm's law. As will be shown in Section 2.8, the right hand side of this equation is small in the MHD parameter range. For example, the first term, which represents the electron inertia, introduces the scale length of the electromagnetic skin depth c/ω_{pe}, the second term, which represents the electron pressure gradient, introduces the scale of the ion Larmor radius at the electron temperature $\rho_s=\sqrt{T_e/m_i}/\omega_{ci}$, and the third term represents the finite frequency effect, ω/ω_{ci}, all of which are regarded as small parameters in the MHD scale. Hence, in the ideal MHD, Eq. (1.6.14) assumes the form called Ohm's law,

$$\mathbf{E}+\mathbf{v}\times\mathbf{B}=0. \tag{1.6.15}$$

This equation shows that the electric field parallel to $\mathbf{B}$ is zero. Unlike the rest of the MHD equations, Eq. (1.6.15) is somewhat strange in that it does not represent a causal relation. Rather, the equation can be regarded as one which either gives **E** or one which gives **v**.

If we combine Eqs. (1.6.10) and (1.6.15) with the adiabatic law (1.6.6) and Maxwell's equations, we have the complete set of ideal magnetohydrodynamic equations,

$$m_i n \frac{d\mathbf{v}}{dt} = \mathbf{J} \times \mathbf{B} - \nabla p \tag{1.6.10}$$

$$\mathbf{E} + \mathbf{v} \times \mathbf{B} = 0 \tag{1.6.15}$$

$$\frac{\partial n}{dt} + \nabla \cdot (n\mathbf{v}) = 0 \tag{1.6.2}$$

$$\nabla \times \mathbf{E} = -\frac{\partial \mathbf{B}}{\partial t} \tag{1.2.5}$$

$$\nabla \times \mathbf{B} = \mu_o \mathbf{J} \tag{1.6.16}$$

$$\nabla \cdot \mathbf{B} = 0\,. \tag{1.2.6}$$

$$\frac{d\ell np}{dt} = \gamma \frac{d\ell nn}{dt}; \quad \gamma = \frac{5}{3}, \quad p = nT\,. \tag{1.6.6}$$

Since there arises little net charge separation in an MHD plasma, the displacement current in Eq. (1.2.7) is ignored here and is replaced by Eq. (1.6.16).

When we apply MHD equations to space plasmas we should keep in mind the assumption of isotropic pressure in (1.6.10), the absence of the parallel electric field in (1.6.15) and, in addition, the low frequency and long wavelength assumptions. These effects will be discussed in later sections.

Energy Density Conservation in MHD Equations

The energy conservation derived from the second moment of the Vlasov equation in Section 1.6 contains cubic moments which were left as unknown. The adiabatic relation for the pressure variations assumed in MHD can close the second moment equation and the energy conservation law can be derived using only known variables.

The energy conservation law can be derived by constructing a scalar product of **v** and Eq. (1.6.10).

$$m_i n\mathbf{v} \cdot \frac{d\mathbf{v}}{dt} = \mathbf{v} \cdot (\mathbf{J} \times \mathbf{B}) - \mathbf{v} \cdot \nabla p \tag{1.6.17}$$

the left hand side becomes,

$$m_i n\mathbf{v}\cdot\frac{d\mathbf{v}}{dt}=m_i n\mathbf{v}\cdot\left[\frac{\partial\mathbf{v}}{dt}+\mathbf{v}\cdot(\nabla\mathbf{v})\right]$$

$$=m_i n\left[\frac{\partial}{\partial t}(\frac{v^2}{2})+\mathbf{v}\cdot\{\nabla\frac{v^2}{2}-\mathbf{v}\times(\nabla\times\mathbf{v})\}\right]$$

$$=\frac{\partial}{\partial t}(\frac{m_i n v^2}{2})-\frac{m_i v^2}{2}\frac{\partial n}{\partial t}+\nabla\cdot\left(\frac{m_i n v^2\mathbf{v}}{2}\right)$$

$$-\frac{m_i v^2}{2}\nabla\cdot(n\mathbf{v}).$$

Using the mass conservation law (1.6.2), we have

$$m_i n\mathbf{v}\cdot\frac{d\mathbf{v}}{dt}=\frac{\partial}{\partial t}\left(\frac{m_i n v^2}{2}\right)+\nabla\cdot\left(\frac{m_i n v^2\mathbf{v}}{2}\right). \tag{1.6.18}$$

The first term on the right hand side of Eq. (1.6.17) becomes, with the help of Maxwell's equations (1.2.5) and (1.6.16) and Ohm's law (1.6.15),

$$\mathbf{v}\cdot(\mathbf{J}\times\mathbf{B})=\mathbf{J}\cdot(\mathbf{B}\times\mathbf{v})=\mathbf{J}\cdot\mathbf{E}$$

$$=-\frac{1}{\mu_o}\nabla\cdot(\mathbf{E}\times\mathbf{B})-\frac{\partial}{\partial t}(\frac{B^2}{2\mu_o}). \tag{1.6.19}$$

The second term becomes, with the help of the mass conservation, (1.6.2) and the adiabatic law (1.6.6),

$$\mathbf{v}\cdot\nabla p=\nabla\cdot(\mathbf{v}p)-p\nabla\cdot\mathbf{v}$$

$$=\nabla\cdot(\mathbf{v}p)+p\frac{d}{dt}\ell n\ n$$

$$=\nabla\cdot(\mathbf{v}p)+\frac{p}{\gamma}\frac{d}{dt}\ell n\ p$$

$$=\nabla\cdot(\mathbf{v}p)+\frac{1}{\gamma}\left[\frac{\partial p}{\partial t}+\mathbf{v}\cdot\nabla p\right].$$

Hence

$$\mathbf{v}\cdot\nabla p=\nabla\cdot\left(\frac{\gamma p\mathbf{v}}{\gamma-1}\right)+\frac{\partial}{\partial t}\left(\frac{p}{\gamma-1}\right). \tag{1.6.20}$$

Combining Eqs. (1.6.18) to (1.6.20), we have the equation of energy density conservation for MHD,

$$\frac{\partial}{\partial t}\left(\frac{m_i n v^2}{2}+\frac{B^2}{2\mu_o}+\frac{p}{\gamma-1}\right)+\nabla\cdot\left(\frac{m_i n v^2 \mathbf{v}}{2}+\frac{\mathbf{E}\times\mathbf{B}}{\mu_o}+\frac{\gamma p \mathbf{v}}{\gamma-1}\right)=0. \quad (1.6.21)$$

Here the first term is the kinetic energy associated with the plasma flow, the second term is the magnetic field energy and the third term is the expansion energy under the adiabatic assumption. Since the displacement current is ignored in MHD based on the quasi-neutrality assumption, the electric field energy is absent here. In addition, the adiabatic assumption eliminates the energy flow due to the thermal flow and heat conduction.

Motion of Magnetic Field

In plasma physics we often discuss a motion of the magnetic field. Since there exists no concept of field line motion in the Maxwell equations, such a concept derives from magnetohydrodynamics. For example, consider the earth's magnetic field. It is considered to rotate with the earth. However, if the field is an ideal dipole, the nature of the electromagnetic field is invariant whether the field line is rotating or not. There is no way to distinguish one field line from another. Thus why do we bother whether the field line is moving or not?

The difference between a rotating field and a nonrotating field appears when a charged particle is placed in the field. A charged particle which suddenly appears in a rotating dipole field will be "caught" by the magnetic field and will move with the field line. Hence by observing the motion of the charged particle, we can identify that the magnetic field is moving.

In a perfectly conducting fluid where Ohm's law, Eq. (1.6.15), is satisfied, the magnetic field **B** can be shown to move with the plasma in a plane normal to **B**. Let us take the curl of Ohm's law, Eq. (1.6.15), and substitute the Maxwell equations (1.2.5) and (1.2.6),

$$\frac{\partial \mathbf{B}}{\partial t}+(\mathbf{v}\cdot\nabla)\mathbf{B}+\mathbf{B}\nabla\cdot\mathbf{v}-(\mathbf{B}\cdot\nabla)\mathbf{v}=0. \quad (1.6.22)$$

On a surface normal to the magnetic field **B**, the plasma $\mathbf{E}\times\mathbf{B}$ velocity field $\mathbf{v}_E$ lies in the plane. Thus, if we take a scalar product of the unit vector of the magnetic field **b** and Eq. (1.6.22), we have

$$\frac{\partial B}{\partial t}+\nabla\cdot(\mathbf{v}_E B)=0. \quad (1.6.23)$$

Equation (1.6.23) can be physically interpreted as if the magnetic field moves with the plasma at the $\mathbf{E}\times\mathbf{B}$ velocity. This is called the frozen-in condition of the magnetic field. It should be recognized, however, that the frozen-in condition breaks down when ideal Ohm's law is not satisfied, for example, due to the finite plasma resistivity.

As the solar wind expands from the solar surface, it is considered to carry the magnetic field because of this mechanism. The foot of a field line moves with the solar surface corotating with the sun also because of this mechanism. Hence the solar magnetic field has a spiral structure in the interplanetary space as will be discussed in Section 3.

1.7 Guiding Center Motions

As we saw in Section 1.4, individual plasma particles in a magnetic field exhibit a Larmor motion around the field line. For most cases in space plasmas, the ion Larmor radius is regarded as small compared with the scale size of its inhomogeneity. We furthermore assume that the temporal variation of the fields is much slower than the ion cyclotron frequency. In these cases, the motion of the plasma particles can be traced by following their "guiding center" motion, the center of the Larmor motion. The motion of the guiding center is closely related to the fluid velocity considered in Section 1.6. Hence, the study of the guiding center motion of the individual particle gives a good insight into the fluid behavior of a plasma.

Magnetic Moments, μ

In the guiding center motion, the important quantity associated with the Larmor motion of a particle is the magnetic moment μ. The magnitude of μ is defined as

$$\mu = \text{current} \times \text{area}. \tag{1.7.1a}$$

$$= q \frac{\omega_c}{2\pi} \pi \rho^2 = \frac{w_\perp}{B}.$$

Here, $w_\perp \left(= \frac{1}{2} m v_\perp^2 \right)$ is the perpendicular kinetic energy with $v_\perp = \sqrt{v_x^2 + v_y^2}$ being the perpendicular velocity. As shown in Section 1.4, the magnetic field produced by the Larmor motion is in the direction opposite to the given field (i.e., diamagnetic), the direction of the vector $\boldsymbol{\mu}$ is taken in the $-\mathbf{b}$ direction, where $\mathbf{b}$ $(= \mathbf{B}/B)$ is the unit vector in the direction of the magnetic field.

$$\boldsymbol{\mu} = -\frac{w_\perp}{B} \mathbf{b}. \tag{1.7.1b}$$

An important property of the magnetic moment is that it is conserved with respect to slow variations of electromagnetic fields either in space or in time. To prove this, we recognize that μ is proportional to the magnetic flux ψ_c surrounded by the Larmor motion of the particle because

$$\psi_c = \oint \mathbf{B} \cdot d\mathbf{S} = B\pi\rho^2 = \frac{2\pi m}{q^2}\,\mu . \tag{1.7.2}$$

Hence, if we take the total time derivative of μ

$$\frac{d\mu}{dt} = \frac{q^2}{2\pi m}\frac{d}{dt}\oint \mathbf{B} \cdot d\mathbf{S}$$

$$= \frac{q^2}{2\pi m}\left[\oint \frac{\partial \mathbf{B}}{\partial t} \cdot d\mathbf{S} + \mathbf{v} \cdot \nabla \oint \mathbf{B} \cdot d\mathbf{S}\right]$$

$$= \frac{q^2}{2\pi m}\left[-\oint \mathbf{E} \cdot d\mathbf{l} + \mathbf{v} \cdot \nabla \oint \mathbf{A} \cdot d\mathbf{l}\right]. \tag{1.7.3}$$

The first term in Eq. (1.7.3) contributes only when the applied electric field has components in the plane of the Larmor motion *and* the field varies within the Larmor radius. The second term is nonzero only when there exists an inhomogeneity in the magnetic field within the Larmor radius. Alternatively, the variation of μ means a change of magnetic flux within the Larmor radius, indicating the exchange of energy between the field and particle. Hence, μ is often called an adiabatic invariant.

E×B Drift, $\mathbf{v}_E$

Let us study the motion of a particle in the presence of a constant electric field which is perpendicular to the constant magnetic field. The equation of motion for a single particle with velocity $\mathbf{v}(t)$ reads

$$m\frac{d\mathbf{v}}{dt} = q\,(\mathbf{E} + \mathbf{v} \times \mathbf{B}). \tag{1.7.4}$$

We introduce a guiding center velocity $\mathbf{v}_G$ as the difference between $\mathbf{v}$ and the velocity due to the Larmor motion $\mathbf{v}_L$

$$\mathbf{v}_G = \mathbf{v} - \mathbf{v}_L . \tag{1.7.5}$$

Substituting Eq. (1.7.5) to (1.7.4), we see

$$m\left(\frac{d\mathbf{v}_G}{dt} + \frac{d\mathbf{v}_L}{dt}\right) = q\,(\mathbf{E} + \mathbf{v}_G \times \mathbf{B} + \mathbf{v}_L \times \mathbf{B}). \tag{1.7.6}$$

If we take $\mathbf{E} \times \mathbf{v}_G + \mathbf{B} = 0$, $\mathbf{v}_G$ is constant in time, and Eq. (1.7.6) becomes

$$m\frac{d\mathbf{v}_L}{dt} = q(\mathbf{v}_L \times \mathbf{B}), \tag{1.7.7}$$

agreeing with the definition of the Larmor motion. Hence, the guiding center drifts with a constant velocity given by

$$\mathbf{v}_G = \mathbf{v}_E = \frac{\mathbf{E}\times\mathbf{B}}{B^2}. \tag{1.7.8}$$

We note that the $\mathbf{E}\times\mathbf{B}$ drift velocity is independent of the charge and the mass of the particle, hence both electrons and ions drift with the same velocity. Eq. (1.6.15) is the manifestation of this fact, although it should be recognized that the velocity $\mathbf{v}$ in (1.6.15) is an Eulerian variable while that in Eq. (1.7.8) is a Lagrangian variable.

F × B Drift, $\mathbf{v}_f$

In the presence of a constant external force $\mathbf{F}$ (such as the gravity) perpendicular to $\mathbf{B}$, the guiding center also drifts. The drift velocity $\mathbf{v}_F$ is given by replacing $\mathbf{E}$ in Eq. (1.7.8) by $\mathbf{F}/q$

$$\mathbf{v}_F = \frac{\mathbf{F}\times\mathbf{B}}{qB^2}. \tag{1.7.9}$$

Unlike the case of $\mathbf{v}_E$, the sign of $\mathbf{v}_F$ depends on the sign of charge q. Hence in the presence of an external force, electrons and ions drift in opposite directions.

Polarization Drift, $\mathbf{v}_p$

When the applied electric field varies with time, the guiding center drifts in the direction of the electric field. From Eq. (1.7.6), $\mathbf{v}_G$ due to the time varying electric field is obtained, after eliminating $\mathbf{v}_L$ and $\mathbf{v}_E$,

$$\mathbf{v}_G = \frac{-\dfrac{m}{q}\dfrac{d\mathbf{v}_G}{dt}\times\mathbf{B}}{B^2}.$$

If the time rate of change of the electric field is much slower than the cyclotron frequency, $\mathbf{v}_G$ on the right hand side may be approximated by $\mathbf{v}_E$, hence

$$\mathbf{v}_G \equiv \mathbf{v}_p = \frac{-\dfrac{m}{q}\dfrac{d}{dt}\left(\dfrac{\mathbf{E}\times\mathbf{B}}{B^2}\right)\times\mathbf{B}}{B^2}$$

$$= \frac{m}{qB^2}\frac{d\mathbf{E}}{dt}. \tag{1.7.10}$$

Since $\mathbf{v}_P$ is proportional to the mass of the particle, polarization drift is dominated by ions.

Curvature Drift, $\mathbf{v}_R$

When the magnetic field is curved, a particle moving along the magnetic line of force encounters the centrifugal force $\mathbf{F}_R$ given by

$$\mathbf{F}_R = \frac{m v_{\parallel}^2}{R^2}\,\mathbf{R},$$

where $v_{\parallel}$ is the particle velocity parallel to the direction of the local magnetic field, and $\mathbf{R}$ is the radius of curvature. Substituting $\mathbf{F}_R$ into the $\mathbf{F} \times \mathbf{B}$ drift velocity (1.7.9), we have the curvature drift, $\mathbf{v}_R$,

$$\mathbf{v}_R = \frac{2 w_{\parallel}\, \mathbf{R} \times \mathbf{B}}{q R^2 B^2}, \qquad (1.7.11)$$

where $w_{\parallel} \left(= \frac{1}{2}\, m\, v_{\parallel}^2\right)$ is the parallel kinetic energy of the particle. The radius of curvature can be expressed using the unit vector in the direction of the magnetic field $\mathbf{b} = \mathbf{B}/B$, (where $B = \sqrt{\mathbf{B} \cdot \mathbf{B}}$),

$$\frac{\mathbf{R}}{R^2} = -\,(\mathbf{b} \cdot \nabla)\,\mathbf{b}. \qquad (1.7.12)$$

Furthermore, if we use the vector identity,

$$\frac{1}{2}\,\nabla\,(\mathbf{b} \cdot \mathbf{b}) = 0 = \mathbf{b} \times (\nabla \times \mathbf{b}) + (\mathbf{b} \cdot \nabla)\mathbf{b},$$

$$\mathbf{v}_R = \frac{2 w_{\parallel}}{q B^2}\,[\mathbf{B} \times (\mathbf{b} \cdot \nabla)\mathbf{b}]$$

$$= \frac{2 w_{\parallel}}{q B^2}\,\mathbf{B} \cdot \mathbf{b}\,(\nabla \times \mathbf{b})_{\perp}$$

hence

$$\mathbf{v}_R = \frac{2 w_{\parallel}}{q B}\,(\nabla \times \mathbf{b})_{\perp}$$

$$= \frac{v_{\parallel}^2}{\omega_c}\,(\nabla \times \mathbf{b})_{\perp}, \qquad (1.7.13)$$

where subscript $\perp$ indicates the component perpendicular to $\mathbf{B}$.

∇B Drift, v_B

When the magnetic flux density varies in the plane perpendicular to the direction of the magnetic field, a drift velocity is induced in order to preserve the adiabaticity. If we assume the inhomogeneity of the magnetic field to be in the x direction, $\mathbf{B} = B\,\hat{\mathbf{z}}$ may be written

$$\mathbf{B} \simeq \hat{\mathbf{z}} \left[B(x=0) + x \frac{\partial B}{\partial x} \right]. \tag{1.7.14}$$

For the particle energy to be preserved in the course of the Larmor motion, the magnetic flux within the Larmor motion must be preserved, i.e.,

$$\Delta \psi_c = 0 = \Delta \oint \mathbf{B} \cdot d\mathbf{S}\,, \tag{1.7.15}$$

where

$$d\mathbf{S} = dx(dy + v_y\, dt)\hat{\mathbf{z}}$$

and v_y is the drift speed in the y direction. If we substitute Eq. (1.7.14) into the condition of adiabaticity (1.7.15),

$$\oint B\, dx\, v_y\, dt = -\oint \cdot \frac{\partial B}{\partial x}\, x\, dxdy\,.$$

If we note $\oint dt = 2\pi/\omega_c$, while $\int xdy (= -\rho^2)$ is the negative of the area surrounded by Larmor radius, the drift speed v_y is given by

$$v_y = \frac{1}{2B} \frac{\partial B}{\partial x} \rho^2 \omega_c$$

$$= \frac{w_\perp}{qB^2} \frac{\partial B}{\partial x}\,, \tag{1.7.16}$$

where $w_\perp (= \frac{1}{2} m\, v_\perp^2)$ is the perpendicular kinetic energy. The direction of the drift in general is $\mathbf{b} \times \nabla B$. Hence the ∇B drift velocity is given by

$$\mathbf{v}_B = \frac{w_\perp}{q} \nabla \frac{1}{B} \times \mathbf{b}\,. \tag{1.7.17a}$$

In summary, the guiding center drifts consist of the following terms;

$\mathbf{E} \times \mathbf{B}$ drift;

$$\mathbf{v}_E = \frac{\mathbf{E} \times \mathbf{B}}{B^2}\,, \tag{1.7.8}$$

$\mathbf{F} \times \mathbf{B}$ drift;

$$\mathbf{v}_F = \frac{\mathbf{F} \times \mathbf{B}}{qB^2}, \tag{1.7.9}$$

Polarization drift;

$$\mathbf{v}_p = \frac{m}{qB^2}\frac{d\mathbf{E}}{dt}, \tag{1.7.10}$$

Curvature drift;

$$\mathbf{v}_R = \frac{2w_{\parallel}}{qB}(\nabla \times \mathbf{b})_{\perp} \tag{1.7.13}$$

∇B drift;

$$\mathbf{v}_B = \frac{w_{\perp}}{qB^2}\,\mathbf{b} \times \nabla B. \tag{1.7.17a}$$

$-\mu\nabla$B Force

The ∇B drift velocity obtained in Eq. (1.7.17a) can also be expressed as

$$\mathbf{v}_B = \frac{-\mu\nabla_{\perp}B \times \mathbf{B}}{qB^2}. \tag{1.7.17b}$$

This expression, when compared with Eq. (1.7.9), indicates that ∇B drift is a consequence of a potential force, $-\mu\nabla_{\perp}B$. We show here in fact that $-\mu\nabla B$ also appears as a force in the direction parallel to the magnetic field when the magnetic flux density varies in the direction of **B**.

When the flux density varies in the direction of **B**, $\nabla \cdot \mathbf{B} = 0$ requires the existence of a radial component B_r such that

$$\frac{1}{r}\frac{\partial}{\partial r}(rB_r) + \frac{\partial B}{\partial z} = 0. \tag{1.7.18}$$

Hence, a particle doing a Larmor motion around **B** sees the radial magnetic field given by

$$B_r \simeq -\frac{\rho}{2}\frac{\partial B}{\partial z},$$

and faces the Lorentz force in the parallel direction given by

$$F_{\parallel} = q\,v_{\theta}B_r = -\,q\,\frac{v_{\perp}^2}{2\omega_c}\frac{\partial B}{\partial z}$$

$$= -\,\mu\nabla_{\parallel}B, \tag{1.7.19}$$

where $\nabla_{\parallel}$ is the gradient operator in the direction of the magnetic field.

A dipole field near the equator may be approximated by $B_0(1+z^2/z_0^2)$ where z is the coordinate parallel to the magnetic field. Equation (1.7.19) then provides the equation of motion for a particle near the equator,

$$m\frac{d^2z}{dt^2} = -2\mu\frac{B_0}{z_0^2}z. \tag{1.7.20}$$

If the magnetic moment is assumed to be constant, the particle motion is given by a harmonic oscillator,

$$z = a\cos\omega_b t, \tag{1.7.21}$$

$$\omega_b = \sqrt{\frac{2\mu B_0}{m z_0^2}}. \tag{1.7.22}$$

Such a motion is called the bounce motion of a particle.

1.8 Guiding Center Currents

Perpendicular Current, $\mathbf{J}_\perp$

The guiding center drift velocities (other than $\mathbf{E}\times\mathbf{B}$ drift) obtained in Section 1.7 can produce currents when summed over species in a plasma. The perpendicular current densities due to the drifts are:

the curvature current $\mathbf{J}_R$,

$$\mathbf{J}_R = \frac{2n\,w_{\parallel}}{B}(\nabla\times\mathbf{b})_\perp \tag{1.8.1}$$

$$= \frac{p_{\parallel}}{B}(\nabla\times\mathbf{b})_\perp,$$

∇B currents $\mathbf{J}_B$,

$$\mathbf{J}_B = n w_\perp \nabla\frac{1}{B}\times\mathbf{b}$$

$$= p_\perp \nabla\frac{1}{B}\times\mathbf{b}, \tag{1.8.2}$$

polarization current $\mathbf{J}_p$

$$J_p = \frac{m_i n}{B^2} \frac{dE}{dt}, \tag{1.8.3}$$

and the external force current

$$J_F = \frac{nF \times B}{B^2}. \tag{1.8.4}$$

Here

$$p_{\|} = 2nw_{\|} = mn < v_{\|}^2 > \tag{1.8.5}$$

and

$$p_{\perp} = n w_{\perp} = \frac{1}{2} mn < v_{\perp}^2 >$$

$$\left(= mn < v_x^2 > = mn < v_y^2 > \right) \tag{1.8.6}$$

are the parallel and perpendicular plasma pressure.

We note that in Eq. (1.8.3), the total time derivative contains the convective derivative, $(\mathbf{v} \cdot \nabla) \mathbf{E}$, which plays an important role, for example, in plasma turbulence.

When dealing with the current associated with the guiding center motion, it is extremely important to recognize that the current carried by the guiding center drifts does not constitute the total current in a plasma: the current due to the gyromotion of individual particles must also be included.

If we consider a set of loop currents produced by the Larmor motion of individual particles in a locally homogeneous plasma, the net current appears only at the edge of the plasma because the currents between two neighboring orbits cancel inside the plasma (see Fig. 1.4). Hence the net current penetrated by a line element **dl** produced by N sets of current loops in an area **A** each carrying the current I is given by

$$I_N = \oint NI\, \mathbf{A} \cdot \mathbf{dl} = \oint \mathbf{M} \cdot \mathbf{dl}$$

$$= \oint \nabla \times \mathbf{M} \cdot d\mathbf{S},$$

where $\mathbf{M} = NI\mathbf{A}$ is the magnetization. Hence the magnetization current density $\mathbf{J}_M$ is given by

$$\mathbf{J}_M = \nabla \times \mathbf{M}. \tag{1.8.7}$$

If we recall the definition of the magnetic moment μ in (1.7.1b) **M** is given by

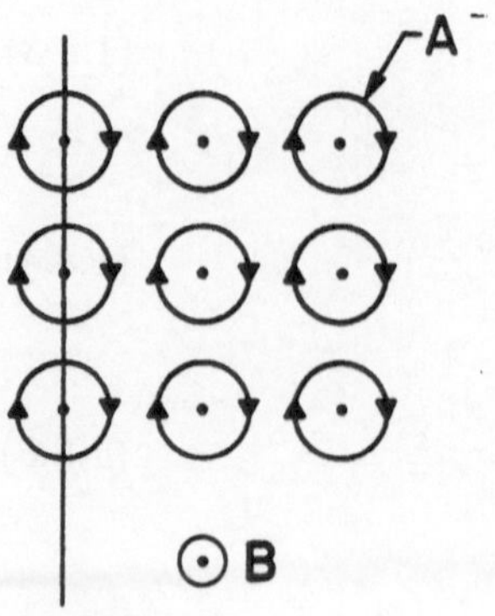

Fig. 1.4. Loop currents to form the magnetization current

$$\mathbf{M} = -\frac{n w_\perp}{B}\,\mathbf{b} = -n\,\boldsymbol{\mu}. \tag{1.8.8}$$

In summary, guiding center currents consist of the following terms, curvature current,

$$\mathbf{J}_R = \frac{2 n w_\parallel}{B}\,(\nabla \times \mathbf{b})_\perp, \tag{1.8.1}$$

∇B current,

$$\mathbf{J}_B = n w_\perp\, \nabla \frac{1}{B} \times \mathbf{b}, \tag{1.8.2}$$

polarization current,

$$\mathbf{J}_p = \frac{m_i n}{B^2}\,\frac{d\mathbf{E}}{dt}, \tag{1.8.3}$$

$\mathbf{F} \times \mathbf{B}$ current,

$$\mathbf{J}_F = \frac{n\mathbf{F} \times \mathbf{B}}{B^2}, \tag{1.8.4}$$

and magnetization current,

$$\mathbf{J}_M = \nabla \times \mathbf{M} = -\nabla \times \left(\frac{n w_\perp}{B}\,\mathbf{b}\right). \tag{1.8.8}$$

Parallel Current $\mathbf{J}_\parallel$

The guiding center currents and magnetization current obtained so far are all directed perpendicular to **B**. What about the parallel (field aligned) current? The parallel current may be obtained from the requirement that the divergence of the total current must vanish,

$$\nabla \cdot \mathbf{J} = \nabla_\perp \cdot \mathbf{J}_\perp + \nabla_\parallel \cdot \mathbf{J}_\parallel = 0, \tag{1.8.9}$$

where $\nabla_\parallel \cdot \mathbf{J}_\parallel = \nabla_\parallel \cdot (J_\parallel \frac{\mathbf{B}}{B}) = \mathbf{B} \cdot \nabla \frac{J_\parallel}{B}$. This equation shows that if $\nabla_\perp \cdot \mathbf{J}_\perp \neq 0$, a parallel current is generated. From the vector identity, $\nabla \cdot \mathbf{J}_M = 0$. Hence the parallel current is generated from the divergence of $\mathbf{J}_R$, $\mathbf{J}_B$ and/or $\mathbf{J}_p$. We note that

$$\nabla_\perp \cdot \mathbf{J}_R = \nabla \cdot \left[\frac{p_\parallel}{B}(\nabla \times \mathbf{b})_\perp\right] \tag{1.8.10}$$

$$\nabla_\perp \cdot \mathbf{J}_B = \nabla_\perp \cdot \left[\nabla\left(\frac{p_\perp}{B}\right) \times \mathbf{b} - \frac{1}{B}\nabla_\perp p_\perp \times \mathbf{b}\right]$$

$$= - \nabla \cdot \left(\frac{p_\perp}{B}(\nabla \times \mathbf{b})_\perp\right) - \nabla_\perp \cdot \left(\frac{\nabla_\perp p_\perp \times \mathbf{b}}{B}\right).$$

Hence

$$\nabla_\perp \cdot \mathbf{J}_R + \nabla_\perp \cdot \mathbf{J}_B = \nabla \cdot \left(\frac{p_\parallel - p_\perp}{B}(\nabla \times \mathbf{b})_\perp\right)$$

$$- \nabla_\perp p_\perp \cdot \nabla_\perp \times \frac{\mathbf{b}}{B}. \tag{1.8.11}$$

Equation (1.8.11) shows that the parallel current may be generated if a pressure gradient exists in the direction of $\mathbf{v}_R$ or $\mathbf{v}_B$. We note that if the pressure is isotropic such that $p_\perp = p_\parallel \equiv p$,

$$\nabla_\perp \cdot \mathbf{J}_R + \nabla_\perp \cdot \mathbf{J}_B = - \nabla_\perp p \cdot \nabla_\perp \times \frac{\mathbf{b}}{B}. \tag{1.8.12}$$

This result can also be obtained from the divergence of the diamagnetic current $\mathbf{J}_D$ of the MHD equation, (1.6.10),

$$\mathbf{J}_D = \frac{-\nabla p \times \mathbf{b}}{B}. \tag{1.8.13}$$

In addition, the divergence of the polarization current (1.8.3) can produce the field aligned current. In particular in a homogeneous plasma

$$\nabla_\perp \cdot \mathbf{J}_p = \nabla_\perp \cdot \left(\frac{m_i n}{B^2} \frac{\partial \mathbf{E}}{\partial t}\right)$$

$$= \frac{m_i n}{B^2} \frac{\partial}{\partial t} \nabla_\perp \cdot \mathbf{E}$$

$$= \frac{m_i n}{B} \mathbf{b} \cdot \frac{\partial}{\partial t} \nabla_\perp \times \mathbf{v}$$

$$= \frac{m_i n}{B} \frac{\partial \Omega}{\partial t}, \tag{1.8.14}$$

where Ω is the vorticity in the direction of $\mathbf{b}$, $\Omega = \mathbf{b} \cdot (\nabla \times \mathbf{v})$, and use is made of Ohm's law. Eq. (1.8.14) indicates that a field aligned current can also be generated in the presence of a time varying vorticity.

In summary, $J_\parallel$ is obtained from the integral along the field line, ℓ,

$$J_\parallel = -B \int \nabla_\perp \cdot \mathbf{J}_\perp d\ell / B$$

$$= B \int [\nabla_\perp p \cdot (\nabla \times \frac{\mathbf{b}}{B}) - \frac{m_i n}{B} \frac{\partial \Omega}{\partial t}] \frac{d\ell}{B}. \tag{1.8.15}$$

1.9 Drift Kinetic Equation

If we use the guiding center velocity obtained in Section 1.8 we can derive a heuristic Vlasov equation for the phase space density in guiding center coordinates, $\mathbf{R}$ ($= X\hat{\mathbf{x}} + Y\hat{\mathbf{y}} + Z\hat{\mathbf{z}}$) and the parallel velocity, $\mathbf{v}_\parallel$. This equation, called the drift kinetic equation, is easier to solve than the Vlasov equation because it allows only one velocity component, $\mathbf{v}_\parallel$. The drift kinetic equation reads

$$\frac{\partial f}{\partial t} + v_\parallel \frac{\partial f}{\partial Z} + \nabla \cdot (\mathbf{v}_d f) + \frac{1}{m} \mathbf{F}_\parallel \cdot \frac{\partial f}{\partial \mathbf{v}_\parallel} = 0 . \tag{1.9.1}$$

Here f is a function of $\mathbf{R}$, $\mathbf{v}_\parallel$, μ and t,

$$\mathbf{F}_\parallel = q\mathbf{E}_\parallel - \mu \nabla_\parallel B \tag{1.9.2}$$

is the parallel force and $\mathbf{v}_d$ is the guiding center velocity derived in Section 1.8. The expression $\mathbf{v}_d$ for electrons and ions are different because of the small mass ratio. The drift velocity to be used for the ion drift kinetic equation is

$$\mathbf{v}_{di} = \frac{\mathbf{E} \times \mathbf{B}}{B^2} + \frac{m_i}{eB^2} \frac{d\mathbf{E}}{dt} + \frac{m_i v_\parallel^2}{eB} (\nabla \times \mathbf{b})_\perp$$

$$+\frac{\mu}{eB}\mathbf{b}\times\nabla_\perp B+v_\parallel\frac{\mathbf{B}_\perp}{B}, \tag{1.9.3}$$

while the drift velocity to be used for the electron drift kinetic equation is

$$\mathbf{v}_{de}=\frac{\mathbf{E}\times\mathbf{B}}{B^2}-\frac{mv_\parallel^2}{eB}(\nabla\times\mathbf{b})_\perp$$

$$-\frac{\mu}{eB}\mathbf{b}\times\nabla_\perp B+v_\parallel\frac{\mathbf{B}_\perp}{B}. \tag{1.9.4}$$

We note that the polarization drift, the second term in Eq. (1.9.3), is important for ions because of the large mass, while the drift due to the bending of the line of force, the last term in Eq. (1.9.4), is important for electrons because of the large parallel velocity. The latter term originates from the $\mathbf{E}\times\mathbf{B}$ drift where $\mathbf{E}$ is given by $\mathbf{v}_\parallel\times\mathbf{B}_\perp/B$.

1.10 Hydromagnetic Equilibrium

It is often found that a plasma in space has such a large energy content that its energy density p is comparable to the magnetic energy density $B^2/2\mu_o$. The ratio of the plasma pressure to the magnetic energy density is often called "beta" and is defined as

$$\beta=\frac{p}{B^2/2\mu_o}=\frac{(T_i+T_e)n}{B^2/2\mu_o}. \tag{1.10.1}$$

When the plasma beta is comparable to unity, the externally given magnetic field is significantly deformed by the "internal" plasma current. In order to obtain the "deformed" magnetic field, one must solve Ampere's law including the internal plasma current; however, the plasma current depends on the magnetic field. Hence, finding a "self-consistent" magnetic field configuration becomes, in general, a nonlinear problem.

Equilibrium with Isotropic Pressure

Let us first approach the problem of equilibrium using the guiding center current obtained in Section 1.9. In a stationary plasma without an external force (other than the magnetic field), the internal currents are given by the magnetization current $\mathbf{J}_M$, the curvature current $\mathbf{J}_R$ and the ∇B current $\mathbf{J}_B$. For simplicity, let us first assume that the plasma has an isotropic Maxwellian distribution. Then $\mathbf{J}_R$ and $\mathbf{J}_B$ are given by

$$\mathbf{J}_R=\frac{p}{B}(\nabla\times\mathbf{b})_\perp \tag{1.10.2}$$

$$\mathbf{J}_B = p\,\nabla\frac{1}{B}\times\mathbf{b}\,. \tag{1.10.3}$$

On the other hand the magnetization current is given by

$$\mathbf{J}_M = -\nabla\times\left(\frac{p}{B}\,\mathbf{b}\right)_\perp$$

$$= -\nabla p\times\frac{\mathbf{b}}{B} - p\nabla\frac{1}{B}\times\mathbf{b} - \frac{p}{B}(\nabla\times\mathbf{b})_\perp\,. \tag{1.10.4}$$

Hence, when $\mathbf{J}_R$, $\mathbf{J}_B$ and $\mathbf{J}_M$ are added together, $\mathbf{J}_R$ cancels with the second term of $\mathbf{J}_M$ and $\mathbf{J}_B$ cancels with the third term of $\mathbf{J}_M$ leaving only the first term in $\mathbf{J}_M$, thus

$$\mathbf{J} = \mathbf{J}_M + \mathbf{J}_R + \mathbf{J}_B = -\nabla p\times\frac{\mathbf{b}}{B}\,. \tag{1.10.5}$$

This result shows that the total current is given by the diamagnetic current. It also shows that the perpendicular current *does not* flow in the direction indicated by ∇B or the curvature drifts of particles but flows in the direction of $\mathbf{B}\times\nabla p$, which depends on the direction of the gradient of the total plasma pressure. This fact is particularly important in considering the current in the magnetosphere or in the tail.

It should be recognized that the current density obtained in (1.10.5) is identical to the equilibrium current of the MHD equation (1.6.10). In the absence of plasma flow, Eq. (1.6.10) gives the equilibrium (force balance) condition,

$$\mathbf{J}\times\mathbf{B} = \nabla p\,. \tag{1.10.6}$$

Solving for $\mathbf{J}$ from Eq. (1.10.6), we can obtain (1.10.5). It is interesting to note that by equating nw (the number density times the energy of the individual particles) to the plasma pressure, the discrete particle picture of the guiding center drift gives a result identical to the equilibrium equation of MHD. Finally, for self-consistency, $\mathbf{J}$ should satisfy Ampere's law

$$\nabla\times\mathbf{B} = \mu_o\mathbf{J}\,. \tag{1.6.16}$$

Equilibrium with Anisotropic Pressure

A useful result of considering the guiding center currents is to distinguish the effect of perpendicular and parallel kinetic energy. Let us now obtain the equilibrium condition assuming a bi-Maxwellian distribution such that

$$n w_\perp (= p_\perp) \neq 2 n w_\parallel (= p_\parallel).$$

Then

$$\mathbf{J}_R = \frac{p_\parallel}{B} (\nabla \times \mathbf{b})_\perp \tag{1.10.7}$$

$$\mathbf{J}_B = p_\perp \nabla \frac{1}{B} \times \mathbf{b}, \tag{1.10.8}$$

while

$$\mathbf{J}_M = -\nabla \times \left(\frac{p_\perp}{B} \mathbf{b} \right)$$

$$= -\nabla p_\perp \times \frac{\mathbf{b}}{B} - p_\perp \nabla \frac{1}{B} \times \mathbf{b} - \frac{p_\perp}{B} (\nabla \times \mathbf{b})_\perp, \tag{1.10.9}$$

and the total current is given by

$$\mathbf{J} = \mathbf{J}_R + \mathbf{J}_B + \mathbf{J}_M .$$

$$= -\nabla p_\perp \times \frac{\mathbf{b}}{B} + \frac{p_\parallel - p_\perp}{B} (\nabla \times \mathbf{b})_\perp . \tag{1.10.10}$$

If we obtain the $\mathbf{J} \times \mathbf{B}$ force from (1.10.10),

$$\mathbf{J} \times \mathbf{B} = -\left(\nabla p_\perp \times \frac{\mathbf{b}}{B} \right) \times \mathbf{B} + \frac{p_\parallel - p_\perp}{B} (\nabla \times \mathbf{b}) \times \mathbf{B}$$

$$= \nabla p_\perp - \mathbf{b}(\mathbf{b} \cdot \nabla p_\perp) + (p_\parallel - p_\perp)(\mathbf{b} \cdot \nabla)\mathbf{b}$$

$$= \nabla_\perp p_\perp + (p_\parallel - p_\perp)(\mathbf{b} \cdot \nabla)\mathbf{b}, \tag{1.10.11}$$

where use is made of the relation

$$\mathbf{b} \times (\nabla \times \mathbf{b}) + (\mathbf{b} \cdot \nabla)\mathbf{b} = 0$$

and

$$\mathbf{b}(\mathbf{b} \cdot \nabla) = \nabla_\parallel , \tag{1.10.12}$$

and $\nabla_\perp$ is the perpendicular component of the gradient. If we now define the pressure tensor $\overleftrightarrow{\mathbf{p}}$ for the anisotropic pressure considered here,

$$\overleftrightarrow{\mathbf{p}} = p_\perp \overleftrightarrow{\mathbf{I}} + (p_\parallel - p_\perp)\, \mathbf{bb}$$

$$= p_\perp \begin{pmatrix} 1 & 0 & 0 \\ 0 & 1 & 0 \\ 0 & 0 & 1 \end{pmatrix} + (p_\parallel - p_\perp) \begin{pmatrix} 0 & 0 & 0 \\ 0 & 0 & 0 \\ 0 & 0 & 1 \end{pmatrix}, \tag{1.10.13}$$

the right hand side of Eq. (1.10.11) can be identified as the perpendicular component of $\nabla \cdot \overleftrightarrow{\mathbf{p}}$, $(\nabla \cdot \overleftrightarrow{\mathbf{p}})_\perp$. If we take divergence of Eq. (1.10.13),

$$\begin{aligned} \nabla \cdot \overleftrightarrow{\mathbf{p}} &= \nabla p_\perp + \mathbf{b}(\mathbf{b} \cdot \nabla)(p_\parallel - p_\perp) + (p_\parallel - p_\perp)\nabla \cdot (\mathbf{bb}) \\ &= \nabla p_\perp + (p_\parallel - p_\perp)(\mathbf{b} \cdot \nabla)\mathbf{b} + (p_\parallel - p_\perp)\mathbf{b}(\nabla \cdot \mathbf{b}) + \mathbf{b}(\mathbf{b} \cdot \nabla)(p_\parallel - p_\perp) \\ &= \nabla_\perp p_\perp + (p_\parallel - p_\perp)(\mathbf{b} \cdot \nabla)\mathbf{b} + \nabla_\parallel p_\parallel + (p_\parallel - p_\perp)\mathbf{b}(\nabla \cdot \mathbf{b}), \end{aligned} \tag{1.10.14}$$

where use is made of the identity

$$\nabla \cdot (\mathbf{bb}) = (\mathbf{b} \cdot \nabla)\mathbf{b} + \mathbf{b}(\nabla \cdot \mathbf{b}).$$

Equation (1.10.14) indicates that the divergence of the pressure tensor $\overleftrightarrow{\mathbf{p}}$ has components both perpendicular and parallel to the magnetic field. The perpendicular component equals $\mathbf{J} \times \mathbf{B}$ derived in Eq. (1.10.11),

$$(\nabla \cdot \overleftrightarrow{\mathbf{p}})_\perp = \nabla_\perp p_\perp + (p_\parallel - p_\perp)(\mathbf{b} \cdot \nabla)\mathbf{b} \tag{1.10.15}$$

while the parallel component is given by

$$(\nabla \cdot \overleftrightarrow{\mathbf{p}})_\parallel = \nabla_\parallel p_\parallel + (p_\parallel - p_\perp)\mathbf{b}(\nabla \cdot \mathbf{b}). \tag{1.10.16}$$

The equilibrium conditions for an anisotropic plasma are hence given by

$$\mathbf{J} \times \mathbf{B} = (\nabla \cdot \overleftrightarrow{\mathbf{p}})_\perp = \nabla_\perp p_\perp + (p_\parallel - p_\perp)(\mathbf{b} \cdot \nabla)\mathbf{b} \tag{1.10.17}$$

and

$$0 = (\nabla \cdot \overleftrightarrow{\mathbf{p}})_\parallel = \nabla_\parallel p_\parallel + (p_\parallel - p_\perp)\mathbf{b}(\nabla \cdot \mathbf{b}). \tag{1.10.18}$$

Flux Function ψ

The equilibrium magnetic field configuration is obtained from

$$\nabla \times \mathbf{B} = \mu_0 \mathbf{J} \tag{1.6.16}$$

$$\nabla \cdot \mathbf{B} = 0, \tag{1.2.6}$$

where $\mathbf{J}$ is given by

$$\mathbf{J} \times \mathbf{B} = \nabla p, \tag{1.10.6}$$

for an isotropic plasma and

$$\mathbf{J} \times \mathbf{B} = \nabla_\perp p_\perp + (p_\parallel - p_\perp)(\mathbf{b} \cdot \nabla)\mathbf{b}, \tag{1.10.11}$$

for an anisotropic plasma. There exists no unique solution for this set of equations since p can have an arbitrary profile. However, for a symmetric case, one can derive a relatively simple scalar field equation for a prescribed pressure profile.

When the field is symmetric either with respect to an axis or to a plane, the geometry contains one ignorable coordinate and the field can be described by a two dimensional coordinate space. The method prescribed here applies to these cases. Let us take an example of axial symmetry with the azimuthal coordinate (the ignorable coordinate) given by ϕ. The divergence-free condition for **B**, Eq. (1.2.6) in general allows one to write **B** in terms of the two potentials α and β, called the Euler potentials,

$$\mathbf{B} = \nabla\alpha \times \nabla\beta \tag{1.10.19}$$

because $\nabla \cdot (\nabla\alpha \times \nabla\beta)$ is identically zero. Equation (1.10.19) means **B** is perpendicular to both the constant α and constant β planes, meaning **B** can be written as a function of α and β. We take β as the ϕ coordinate, then from the vector identity the surface integral gives

$$\psi = \int \mathbf{B} \cdot d\mathbf{S} = \int (\nabla\alpha \times \nabla\beta) \cdot d\mathbf{S} = \oint \alpha \, d\beta$$

$$= \oint \alpha \, d\phi .$$

If we take the surface to be the circular surface whose edge is tangential to the ϕ coordinate this integral gives $2\pi\alpha$. Hence, $2\pi\alpha$ gives the magnetic flux ψ which penetrates the circular plane whose center coincides with the axis of symmetry. Thus in the case of axial symmetry, $\psi = 2\pi\alpha$ and **B** is given by

$$\mathbf{B} = \frac{1}{2\pi} \nabla\psi \times \nabla\phi . \tag{1.10.20}$$

In general **B** can also have a component in the direction of the ignorable coordinate ϕ, which cannot be described by Eq. (1.10.20) but can still be axis symmetric. Since such a case is rare in space plasmas, we ignore the ϕ component of **B** here. We note that in the cylindrical coordinate

$$\nabla\phi = \frac{1}{r}\hat{\phi} \tag{1.10.21}$$

while in the spherical coordinate

$$\nabla\phi = \frac{1}{r\sin\theta}\,\hat{\boldsymbol{\phi}}\,, \tag{1.10.22}$$

where $\hat{\boldsymbol{\phi}}$ is the unit vector in the ϕ direction and θ is the polar angle. The flux function is related to the ϕ component of the magnetic vector potential A_ϕ through

$$\psi = \oint \mathbf{B}\cdot d\mathbf{S} = \oint \mathbf{A}\cdot \mathbf{dl}$$
$$= \begin{cases} 2\pi r A_\phi & \text{for the cylindrical coordinate} \\ 2\pi r\sin\theta A_\phi & \text{for the spherical coordinate}. \end{cases} \tag{1.10.23}$$

It is easily seen that for a case with axial symmetry

$$\mathbf{B} = \nabla\times\mathbf{A} = \nabla\times A_\phi\hat{\boldsymbol{\phi}}$$
$$= \nabla A_\phi \times \hat{\boldsymbol{\phi}}\,, \tag{1.10.24}$$

which is consistent with Eqs. (1.10.20) and (1.10.23).

Grad-Shafranov Equation — Isotropic Pressure

If the plasma pressure is isotropic, Eq. (1.10.6) gives

$$\mathbf{B}\cdot\nabla p = 0\,. \tag{1.10.25}$$

From Eq. (1.10.20), this expression indicates that ∇p is parallel to $\nabla\psi$, that is, p is a function of ψ only,

$$p = p(\psi)\,, \tag{1.10.26}$$

and

$$\nabla p = \frac{dp}{d\psi}\nabla\psi\,. \tag{1.10.27}$$

On the other hand from Eq. (1.10.20),

$$\mathbf{J}\times\mathbf{B} = \frac{1}{2\pi}\,\mathbf{J}\times(\nabla\psi\times\nabla\phi)$$
$$= \frac{1}{2\pi}\Big[(\mathbf{J}\cdot\nabla\phi)\,\nabla\psi - (\mathbf{J}\cdot\nabla\psi)\,\nabla\phi\Big]$$
$$= \frac{1}{2\pi}\,(\mathbf{J}\cdot\nabla\phi)\,\nabla\psi\,. \tag{1.10.28}$$

(Since ∇p has no component in the direction of $\hat{\boldsymbol{\phi}}$, $\mathbf{J}\cdot\nabla\psi = 0$). From (1.10.27) and (1.10.28), we have

$$\mathbf{J}\cdot\nabla\phi = 2\pi\,\frac{dp}{d\psi}\,. \tag{1.10.29}$$

Here, $\mathbf{J} \cdot \nabla\phi$ ($= J_\phi/r$ for the cylindrical coordinates, $= J_\phi/2\pi r\sin\theta$ for the spherical coordinate) can be expressed in terms of ψ using Eqs. (1.6.16) and (1.10.20). For example, in a spherical coordinate, Eq. (1.10.29) reads,

$$\frac{\partial^2\psi}{\partial r^2}+\frac{1}{r^2}\frac{\partial^2\psi}{\partial\theta^2}-\frac{\cot\theta}{r^2}\frac{\partial\psi}{\partial\theta} = -4\pi^2 r^2\sin^2\theta\mu_0\frac{dp}{d\psi}. \tag{1.10.30}$$

This is the expression of the Grad-Shafranov equation for the spherical coordinate. When $p(\psi)$ is prescribed, Eq. (1.10.30) is in principle solvable for ψ, and the corresponding magnetic field which is given by Eq. (1.10.20) automatically satisfies the equilibrium condition. When the plasma is absent, $dp/d\psi = 0$ and the (earth) dipole field, which is given by

$$\psi_D = -\frac{2\pi M\sin^2\theta}{r}, \tag{1.10.31}$$

satisfies this equation as it should.

Grad-Shafranov Equation-Anisotropic Pressure

In space plasmas, anisotropic pressure is quite commonly observed. In such cases p is no longer a function of ψ only, but must also be treated as a function of ψ and $B = \sqrt{\mathbf{B}\cdot\mathbf{B}}$. Even in this case we can derive the Grad-Shafranov equation.

We first note $\nabla\cdot\mathbf{b} = -\dfrac{\mathbf{b}\cdot\nabla B}{B}$. Hence, the parallel pressure balance equation (1.10.18) reads

$$\mathbf{b}\cdot\nabla\, p_\parallel - \frac{p_\parallel - p_\perp}{B}\,\mathbf{b}\cdot\nabla B = 0$$

or

$$\frac{\partial p_\parallel}{\partial B} = \frac{p_\parallel - p_\perp}{B}. \tag{1.10.32}$$

Next, $(\mathbf{b}\cdot\nabla)\mathbf{b}$ in the perpendicular component of $(\nabla\cdot\overleftrightarrow{\mathbf{p}})$ in Eq. (1.10.15) can be written as

$$(\mathbf{b}\cdot\nabla)\mathbf{b} = \frac{1}{B^2}(\mathbf{B}\cdot\nabla)\mathbf{B} - \frac{\mathbf{b}}{B}(\mathbf{b}\cdot\nabla)B,$$

while from

$$\frac{1}{2}\nabla(\mathbf{B}\cdot\mathbf{B}) = \mathbf{B}\times(\nabla\times\mathbf{B}) + (\mathbf{B}\cdot\nabla)\mathbf{B},$$

we have

$$(\mathbf{b}\cdot\nabla)\mathbf{b} = \frac{1}{B}\nabla_\perp B + \frac{(\nabla\times\mathbf{B})\times\mathbf{B}}{B^2}. \tag{1.10.33}$$

Substituting Eq. (1.10.33) into the perpendicular equilibrium relation,

$$\mathbf{J}\times\mathbf{B}=(\nabla\cdot\overleftrightarrow{\mathbf{p}})_\perp$$

$$=\nabla_\perp p_\perp+\frac{p_\parallel-p_\perp}{B}\left[\nabla_\perp B+\frac{(\nabla\times\mathbf{B})\times\mathbf{B}}{B}\right]. \qquad (1.10.34)$$

Substituting Ampere's law $\mathbf{J} = (\nabla \times \mathbf{B})/\mu_0$, into Eq. (1.11.34) we have

$$\frac{\sigma}{\mu_0}(\nabla\times\mathbf{B})\times\mathbf{B}=\nabla_\perp p_\perp+\frac{p_\parallel-p_\perp}{B}\nabla_\perp B,$$

where

$$\sigma = 1-\frac{\mu_0}{B^2}\frac{p_\parallel-p_\perp}{B^2}. \qquad (1.10.35)$$

We further note that

$$\mathbf{B}\times(\nabla\times\sigma\mathbf{B})=\sigma\mathbf{B}\times(\nabla\times\mathbf{B})+\mathbf{B}\times(\nabla\sigma\times\mathbf{B})$$

$$=\sigma\,\mathbf{B}\times(\nabla\times\mathbf{B})+B^2\nabla_\perp\sigma$$

$$=\sigma\,\mathbf{B}\times(\nabla\times\mathbf{B})-\mu_0\,B^2\,\nabla_\perp\,\frac{p_\parallel-p_\perp}{B^2}. \qquad (1.10.36)$$

If we substitute Eq. (1.10.36)) into (1.10.35), we have

$$\frac{1}{\mu_0}(\nabla\times\sigma\mathbf{B})\times\mathbf{B}=\nabla_\perp p_\parallel-\frac{p_\parallel-p_\perp}{B}\nabla_\perp B \qquad (1.10.37)$$

Here we note that $\nabla p_\parallel = \nabla p_\parallel\,(\psi, B)$ gives

$$\nabla_\perp p_\parallel=\frac{\partial p_\parallel}{\partial\psi}\nabla_\perp\psi+\frac{\partial p_\parallel}{\partial B}\nabla_\perp B,$$

and substituting Eq. (1.10.32), we have

$$\nabla_\perp p_\parallel=\frac{\partial p_\parallel}{\partial\psi}\nabla_\perp\psi+\frac{p_\parallel-p_\perp}{B}\nabla_\perp B.$$

If we use this expression in Eq. (1.10.37), the second term on the right hand side cancels and we finally have

$$\frac{1}{\mu_0}(\nabla\times\sigma\mathbf{B})\times\mathbf{B}=\frac{\partial p_\parallel}{\partial\psi}\nabla_\perp\psi. \qquad (1.10.38)$$

This expression is formally analogous to the equilibrium expression for the isotropic pressure case. If we use the flux function expression for B, Eq. (1.10.38) becomes, for example, for a spherical coordinate

$$\frac{\partial^2 \psi}{\partial r^2} + \frac{1}{r^2}\frac{\partial^2 \psi}{\partial \theta^2} - \frac{\cot\theta}{r^2}\frac{\partial \psi}{\partial \theta} + \nabla\psi \cdot \nabla \ell n\sigma$$

$$= -\frac{4\pi^2 r^2 \sin^2\theta}{\sigma}\,\mu_0\,\frac{\partial p_\parallel}{\partial \psi}, \qquad (1.10.39)$$

where σ is given by Eq. (1.10.35). We note that in this case the parallel pressure balance equation should also be satisfied,

$$\mathbf{b} \cdot \nabla\left(\frac{p_\parallel}{B}\right) + \frac{p_\perp}{B^2}\,\mathbf{b} \cdot \nabla B = 0. \qquad (1.10.40)$$

Although in principle Eqs. (1.10.39) and (1.10.40) are solvable given $p_\parallel$ and $p_\perp$ as functions of ψ and B, in reality these quantities are given as functions of coordinates. One way to resolve this problem is to use a perturbation method around the vacuum field. For example, for a vacuum magnetosphere ψ and B are given as explicit functions of r and θ through Eqs. (1.10.31) and (1.10.20). If we write the true flux function ψ in terms of the vacuum field ψ_D and the perturbation around it ψ_1,

$$\psi = \psi_D + \psi_1$$

$$= \frac{M \sin^2\theta}{2r} + \psi_1, \qquad (1.10.41)$$

the observed pressure, say as a function of r, can be fit to ψ_D, and thereby $\partial p_\parallel/\partial\psi$ can be computed approximately as $\partial p_\parallel/\partial\psi_D$ as a function of r and θ. This, as well as $p_\perp$ and $p_\parallel$ as functions of r and θ (through ψ_D and B_0), can be used to solve for ψ_1 in Eq. (1.10.39) knowing that ψ_D satisfies

$$\frac{\partial^2 \psi_D}{\partial r^2} + \frac{1}{r^2}\frac{\partial^2 \psi_D}{\partial \theta^2} - \frac{\cot\theta}{r^2}\frac{\partial \psi_D}{\partial \theta} = 0.$$

The exact equilibrium may be obtained by repeating these processes systematically.

We note that the equilibrium obtained this way does not necessarily warrant its stability. The equilibrium contains free energy in anisotropy and in the localization of the pressure, either of which may be the cause of an instability.

Equilibrium Based on Pitch Angle Distribution

Particle data obtained from a spacecraft often contain information of the particle distribution function in pitch angle θ and energy w. This information, combined with the assumption of conservation of the magnetic moment μ, may be used to derive the variation of plasma pressure $p_{\parallel}$ and $p_{\perp}$ along the magnetic line of force, which is designated by the coordinate ℓ. For the relevant phase space distribution function, it is convenient to use the pitch angle (θ) - energy (w) distribution function $f(w, \theta, 0)$ *at the equator*, so that the number density of the particle at the equator is given by

$$n(0) = \int_0^{\infty} dw \int^{\theta} 2\pi \sin\theta d\theta f(w, \theta, 0). \tag{1.10.42}$$

Here zero indicates the equatorial position along the field line.

When the pitch angle distribution is separable from the energy distribution, one commonly used distribution function is

$$f(w, \theta, 0) = \frac{n_0 \Gamma(s+3/2)}{2\pi(\cos\theta_c)^{2s+1}\Gamma(1/2)\Gamma(s+1)}$$

$$(\sin^2\theta - \sin^2\theta_c)^s f_0(w), \tag{1.10.43}$$

where $f_0(w)$ is the equatorial energy distribution function which is normalized to unity when integrated from $w=0$ to infinity, θ_c is the loss cone angle of the field line which can be assumed to be zero for most portions of the magnetosphere, s is a parameter which designates the degree of anisotropy ($s=0$ corresponds to isotropic distribution), Γ is the gamma function and the angle of integration is $\pi - \theta_c \geq \theta \geq \theta_c$.

Off the equator, $\ell \neq 0$, the number density changes due to the change in the volume element in coordinate space $dV = dSd\ell$ caused by the adiabatic motion of particles on converging field lines, where ℓ is the distance along the field line and S is the cross section of a flux tube. Conservation of J_2 and ψ to be introduced in Section 1.11 requires $dSd\ell v_{\parallel}/B$ to be invariant along the flux tube, where $v_{\parallel}$ is the parallel velocity given, from the energy (w) conservation, by

$$v_{\parallel} = \pm\sqrt{\frac{2}{m}(w - \mu B)}, \tag{1.10.44}$$

while the magnetic moment μ (= const) may be expressed in terms of the equatorial pitch angle,

$$\mu = \frac{w}{B(0)}\sin^2\theta, \tag{1.10.45}$$

where B(0) is the magnetic field at the equator, $\ell = 0$. Consequently, the number density $n(\ell)$ at any point, ℓ, along the field line where the magnetic field is given by $B(\ell)$ is written,

$$n(\ell) = \int_0^{\infty} dw \int^{\theta} 2\pi \sin\theta d\theta \frac{B(\ell)}{B(0)} \frac{v_{\parallel}(0)}{v_{\parallel}(\ell)}$$

$$= \int_0^{\infty} dw \int^{\theta} 2\pi \sin\theta d\theta \frac{B(\ell)}{B(0)} \frac{|\cos\theta| f(w,\theta,0)}{\sqrt{1 - B(\ell)\sin^2\theta / B(0)}}. \qquad (1.10.46)$$

In a similar manner, the parallel and perpendicular pressures, $p_{\parallel} = \langle mv_{\parallel}^2 \rangle = 2\langle w - \mu B \rangle$ and $p_{\perp} = \langle \mu B \rangle$ are given respectively by

$$p_{\parallel}(\ell) = \int_0^{\infty} dw \int^{\theta} 2\pi \sin\theta d\theta \frac{2wB(\ell)}{B(0)} |\cos\theta| [1 - \frac{B(\ell)}{B(0)} \sin^2\theta]^{1/2} f(w,\theta,0) \qquad (1.10.47)$$

and

$$p_{\perp}(\ell) = \int_0^{\infty} dw \int^{\theta} 2\pi \sin\theta d\theta w [\frac{B(\ell)}{B(0)}]^2 \frac{|\cos\theta| \sin^2\theta}{[1 - B(\ell)\sin^2\theta / B(0)]^{1/2}} f(w,\theta,0). \qquad (1.10.48)$$

We note that $p_{\parallel}$ and $p_{\perp}$ obtained here satisfy the parallel pressure balance (1.10.40) as they should since $\mu = \text{const}$ is used in either case. This means that if the equatorial distribution function $f(w, \theta, 0)$ is given from satellite data, the parallel pressure along the field line at a distance ℓ can be obtained from Eq. (1.10.47) while $p_{\perp}(\ell)$ is given by Eq. (1.10.48) or is simply given from Eq. (1.10.32) by

$$p_{\perp}(\ell) = p_{\parallel}(\ell) - B \frac{\partial p_{\parallel}}{\partial B}. \qquad (1.10.49)$$

If we use a dipole field for $B(\ell)$ in these expressions, $p_{\parallel}(\ell)$ and $p_{\perp}(\ell)$ obtained here may be used in Eq. (1.10.39) to obtain the first order correction to the dipole field due to the observed particle distribution function.

1.11 Adiabatic Invariants

When a particle executes a periodic motion, the action J defined as

$$J = \frac{1}{2\pi} \oint P dQ \qquad (1.11.1)$$

is conserved. Here the line integral is taken over a period of the motion in the canonical coordinate Q, and P is the canonical coordinate which is conjugate to Q.

The action J itself can be considered as a canonical variable. In this case the canonical variable which forms the conjugate pair with J is the phase angle θ of the periodic motion. The Hamilton equation of motion then reads

$$\frac{\partial H}{\partial J} = \dot{\theta}(\equiv \Omega) \tag{1.11.2}$$

$$\frac{\partial H}{\partial \theta} = -\dot{J} . \tag{1.11.3}$$

Here Ω is the angular frequency of the periodic motion. When the Hamiltonian is not a function of the phase angle θ, $\partial H / \partial \theta = 0$, hence $J =$ const. Thus the action is conserved in such a case. The Hamiltonian is given by, from Eq. (1.11.2),

$$H = \Omega J . \tag{1.11.4}$$

If the particle has more than one periodic motion, actions can be defined for each of the periodic motions J_1, J_2.. .... If the periodic motions are independent,

$$H = \Omega_1 J_1 + \Omega_2 J_2 + \cdots \tag{1.11.5}$$

A particle in the magnetosphere has three different periodic motions. The first is Larmor motion around earth's magnetic field. The second is the bounce motion along the magnetic line of force. The third is the periodic motion around the earth due to the curvature and gradient B drifts. Although these three motions are coupled, and hence the Hamiltonian cannot be written in an uncoupled form as Eq. (1.11.5), one can define actions for each of these motions in an approximate form. If the earth's magnetic field is assumed to be axially symmetric around the north-south axis, the field can be expressed by the ϕ component of the vector potential $A_\phi(r, \theta)$ in the spherical coordinate. Furthermore if we consider a particle trapped near the equatorial plane, the Hamiltonian Eq. (1.2.13) may be approximated by

$$H = \frac{1}{2m}\left(\frac{P_\phi}{r} - qA_\phi\right)^2 + \frac{1}{2m}P_r^2 + \frac{1}{2m}P_z^2 , \tag{1.11.6}$$

where z is the coordinate in the direction of the magnetic field, P_r and P_z are mv_r and mv_z respectively, and $P_\phi = mr^2\dot{\phi} + qA_\phi r$.

Let us compute actions for a particle moving in this magnetic field. For illustrative purposes, we first compute the action J_2 due to the bounce motion along the field line. From the definition of Eq. (1.11.1),

$$J_2 = \oint P_z dz$$

$$= \oint mv_z \frac{dz}{dt} dt$$

$$= \oint m v_z^2 dt \, . \tag{1.11.7}$$

As was shown in Eq. (1.7.21), if μ is conserved, bounce motion is periodic with the bounce frequency ω_b, v_z is given by $\overline{v}_z \cos \omega_b t$ and J_2 is given by

$$J_2 = \frac{\frac{1}{2} m \overline{v}_z^2}{\omega_b} = \frac{w_\parallel}{\omega_b} \, , \tag{1.11.8}$$

where $w_\parallel (= \frac{1}{2} m \overline{v}_z^2)$ is the kinetic energy associated with the parallel motion. We note that the action has a dimension of energy divided by frequency, the dimension of angular momentum. The Hamiltonian which describes the bounce motion is given by $\mu B + 1/2 m v_z^2$, and if B is parabolic as assumed in Section 1.7, H does not contain the phase angle $\theta_b = \int \omega_b dt$ associated with the bounce motion, thus $\partial H / \partial \theta_b = \dot{J}_2 = 0$. Hence J_2 is a constant of motion. We note, however, that if μ is not conserved, or if there exists a parallel electric field, this is no longer the case. Hence J_2 is conserved in the sense there exists no exchange of energy between the particle kinetic energy and the external field. Analogous to the case of the magnetic moment, μ, J_2 is also considered as an adiabatic invariant. J_2 is called the second adiabatic invariant in order to distinguish it from μ.

Let us now show that the magnetic moment μ is in fact proportional to the action due to the Larmor motion of the particle. Since the Larmor motion is associated with the r components of P, we take

$$\begin{aligned} J_1 &= \frac{1}{2\pi} \oint P_r dQ_r \\ &= \frac{1}{2\pi} \oint m v_r^2 dt \\ &= \frac{m v_\perp^2}{2\omega_c} \, , \end{aligned} \tag{1.11.9}$$

Thus Eq. (1.11.9) gives

$$J_1 = \frac{m v_\perp^2}{2\omega_c} = \frac{m}{q} \mu \, . \tag{1.11.10}$$

The third adiabatic invariant, J_3, originates from the periodic motion associated with the azimuthal (longitudinal) drift. The canonical coordinate which is conjugate to P_ϕ is ϕ, while H is independent of ϕ. Hence $\dot{P}_\phi = -\partial H/\partial\phi = 0$, thus P_ϕ is the constant of motion. J_3 is thus given by

$$J_3 = \frac{1}{2\pi}\oint P_\phi dQ_\phi = mr^2\dot{\phi} + qA_\phi r . \qquad (1.11.11)$$

As is seen in Section 1.7, the drift speed $r\dot{\phi}$ is smaller than qA_ϕ/m by a factor of $(\rho/r)^2$, where ρ is the Larmor radius, hence the first term in Eq. (1.11.11) is negligible and J_3 is approximately given by

$$J_3 \simeq \frac{q}{2\pi}\psi , \qquad (1.11.12)$$

where ψ is the total magnetic flux surrounded by the drift orbit, see Eq. (1.10.23).

The canonical pair of the action and phase angle are called the action-angle variables. If the action-angle variables are used as canonical variables, the stationary distribution function becomes a function of actions, i.e.,

$$f = f(J_1, J_2, J_3) \qquad (1.11.13)$$

because the Poisson bracket, Eq. (1.3.9), vanishes,

$$\{f, H\} = \frac{\partial H}{\partial J_j}\frac{\partial f}{d\theta_j} - \frac{\partial H}{\partial \theta_j}\frac{\partial f}{\partial J_j} = 0 .$$

Hence if the magnetic field is axially symmetric, the stationary distribution function is generally given by

$$f = f(\mu, J_2, \psi) . \qquad (1.11.14)$$

Note that in this coordinate system, the volume element in velocity space is given by

$$d\mathbf{v} = \frac{B}{m^2} d\mu dJ_2/(v_z\oint dz/v_z) \qquad (1.11.15)$$

and in coordinate space the volume element is given by

$$dV = d\,\psi\, d\,\phi\, dz/B . \qquad (1.11.16)$$

1.12 Transport Coefficients

Interparticle collisions allow fluid quantities such as the density, currents, momentum and energy to be transported across the ambient magnetic field. There are a large number of transport coefficients because of anisotropies produced by the magnetic field. Readers are advised to consult the famous article by S. I. Braginskii (*Review of Plasma*, Vol. 1 p. 205) for the complete list of the coefficients. In this section we introduce only a few important coefficients which are frequently used in space plasmas.

Resistivity

Electron-ion collisions produce friction in the electron current. The electric resistivity which arises from these collisions modifies Ohm's law (1.6.15) to

$$\mathbf{E}+\mathbf{v}\times\mathbf{B}=\eta\mathbf{J}, \tag{1.12.1}$$

where the resistivity η is given by

$$\eta=\frac{\nu_{ei}m_e}{e^2n}=\frac{\nu_{ei}}{\epsilon_0\omega_{pe}^2}, \tag{1.12.2}$$

and ν_{ei} is the electron ion collision frequency given in Eq. (1.4.13). The electric conductivity σ_e is given by η^{-1} and

$$\sigma_e=\frac{\epsilon_0\omega_{pe}^2}{\nu_{ei}}. \tag{1.12.3}$$

Particle Diffusion

The presence of the resistivity produces a flow in the direction perpendicular to the current and the magnetic field given from Eq. (1.12.1) by

$$\mathbf{v}_r=\frac{\mathbf{B}\times\eta\mathbf{J}}{B^2}. \tag{1.12.4}$$

If the plasma is in equilibrium, the plasma current is given by the diamagnetic current, Eq. (1.8.13). If we use the diamagnetic current in Eq. (1.12.4), the resistive flow becomes,

$$\mathbf{v}_r=-\frac{\eta\nabla p}{B^2}. \tag{1.12.5}$$

Substitution of Eqs. (1.12.2) and (1.12.5) into the mass conservation equation, (1.6.2) yields,

$$\frac{\partial n}{\partial t}=\nabla_\perp\cdot\left(\frac{\nu_{ei}m_e}{e^2B^2}\nabla_\perp p\right). \tag{1.12.6}$$

If we consider an isothermal situation such that the temperature is uniform in the plasma $\nabla_\perp p=T\nabla_\perp n$, Eq. (1.12.6) gives the diffusion equation in the direction perpendicular to the ambient magnetic field,

$$\frac{\partial n}{\partial t}=\nabla_\perp\cdot(D\nabla_\perp n), \tag{1.12.7}$$

where the (perpendicular) diffusion coefficient is given by

$$D = \frac{\nu_{ei} m_e T}{e^2 B^2} = \nu_{ie} \bar{\rho}_i^2 \tag{1.12.8}$$

and

$$\bar{\rho}_i = \left(\frac{T}{m_i}\right)^{1/2} \frac{1}{\omega_{ci}} \tag{1.12.9}$$

is the ion Larmor radius at the total plasma temperature, $T = T_e + T_i$. Equation (1.12.8) shows that particle diffusion in a fully ionized plasma is given by the movement of the guiding center position by a distance equal to the ion Larmor radius at each ion-electron collision time (or equivalently by a distance equal to the electron Larmor radius ρ_e $(= (T/m_e)^{1/2}/\omega_{ce})$ at each electron-ion collision time). Since the ion-electron collision frequency is smaller than the electron-ion collision frequency by the ion to electron mass ratio, particle diffusion by Coulomb collision is extremely slow. However, when the plasma is turbulent, the correlation time of the density and velocity field fluctuations may become much shorter than ν_{ie}^{-1}. Within the MHD fluctuations, the shortest correlation time would be of the order of ω_{ci}^{-1}. The turbulent diffusion coefficient may then be given by

$$D_T = \alpha \omega_{ci} \rho_i^2$$

$$= \alpha \frac{T}{eB}. \tag{1.12.10}$$

When we take $\alpha = 1/16$ and T to be the electron temperature, the expression for this diffusion coefficient is called the Bohm diffusion coefficient.

Heat Diffusion

The equation for heat diffusion may be obtained from the MHD energy conservation equation, Eq. (1.6.21). If we ignore the energy due to the convection and the magnetic field and assume a uniform density, the heat diffusion equation may be written as

$$\frac{3}{2} n \frac{\partial T}{\partial t} = \nabla_\perp \cdot (\kappa_\perp \nabla_\perp T) + \nabla_\parallel \cdot (\kappa_\parallel \nabla_\parallel T). \tag{1.12.11}$$

Here, unlike the case of density diffusion where the ambipolarity requires an identical diffusive velocity for electrons and ions, the thermal flow velocity for electrons and ions can be different. This leads to a different heat conductivity for each species.

Since electrons are more difficult to move in the direction perpendicular to the magnetic field, ions play the dominant role in the perpendicular heat conductivity. Thus,

$$\kappa_\perp \simeq \kappa_{\perp i} = 2n\rho_i^2 \nu_{ii}, \tag{1.12.12}$$

while electrons play the dominant role in the parallel heat conductivity,

$$\kappa_\parallel \simeq \kappa_{\parallel e} = 3.2 \frac{n v_{Te}^2}{\nu_{ee}}, \tag{1.12.13}$$

where ν_{ii} and ν_{ee} are the ion-ion and electron-electron collision frequencies respectively (see Sect. 1.4).

1.13 Particle Distribution Function

In this section, we consider some specific forms of the particle distribution functions which are pertinent to space plasmas. As was shown in Section 1.3, the solution of the equilibrium Vlasov equation for a particle distribution function can be given by a general function of the Hamiltonian, f(H). Since the Hamiltonian in an electrostatic potential ϕ is given by

$$H = \frac{1}{2} m v^2 + q\phi, \tag{1.13.1}$$

the equilibrium distribution becomes $f(1/2mv^2 + q\phi)$, where q and m are the charge and the mass of the particle and v is the velocity. We also note that, if the plasma is two-dimensional in such a way that there exists an ignorable coordinate, say y, then the y-component of the canonical momentum P_y is also a constant of motion (because $\partial H/\partial y = \dot{P}_y = 0$), and the distribution function becomes also a function of $P_y = mv_y + qA_y$, i.e.,

$$f = f(\frac{1}{2} m v^2 + q\phi, mv_y + qA_y). \tag{1.13.2}$$

Now, what type of specific function should we reasonably assume? Since the Vlasov equation does not give any more information, we must bring in other physics to answer this question. Normally, we consider that inter-particle collisions determine the specific distribution function. In the presence of collisions, the distribution function is believed to reach a form which will maximize the entropy, S, defined by

$$S = -\int f \ln f \, d\mathbf{v}. \tag{1.13.3}$$

Let us first obtain the distribution function which maximizes the entropy without asking about specific collisional processes. To maximize the entropy, we must use the constraints that the number density

$$n = \int f \, d\mathbf{v} \tag{1.13.4}$$

and the energy density

$$E = \int H f \, d\mathbf{v} \tag{1.13.5}$$

are conserved in the process of entropy increase. Then the variation to maximize the entropy gives,

$$-\delta \int f \ell n f \, d\mathbf{v} - \lambda_1 \delta \int f \, d\mathbf{v} - \lambda_2 \delta \int H f \, d\mathbf{v} = 0, \tag{1.13.6}$$

where λ_1 and λ_2 are Lagrangian multipliers. Taking the variation with respect to f, we obtain,

$$\int \delta f (\ell n f + 1 + \lambda_1 + \lambda_2 H) d\mathbf{v} = 0,$$

which gives,

$$f = e^{-1-\lambda_1-\lambda_2 H}. \tag{1.13.7}$$

If we choose λ_1 and λ_2 such that the distribution function is normalized to unity when integrated over $d\mathbf{v}$ at $\phi = 0$, then f gives the Maxwell-Boltzmann distribution,

$$f(H) = \left(\frac{m}{2\pi T}\right)^{3/2} \exp\left[-\frac{mv^2/2 + q\phi}{T}\right]. \tag{1.13.8}$$

Let us now consider a specific collisional process which can lead to the Maxwell distribution. The evolution of the distribution function due to collisions can be described by the Fokker-Planck equation,

$$\frac{\partial f}{\partial t} = \frac{\partial}{\partial \mathbf{v}} \cdot \left[\frac{1}{2} D(\mathbf{v}) \cdot \frac{\partial f}{\partial \mathbf{v}} - \mathbf{v}\nu(\mathbf{v}) f\right], \tag{1.13.9}$$

where the Diffusion tensor (in velocity space) is given by

$$D(\mathbf{v}) = D_{\parallel} \mathbf{I} + D_{\perp}\left(\mathbf{I} - \frac{\mathbf{v}\mathbf{v}}{v^2}\right). \tag{1.13.10}$$

When the collision is through the Coulomb field which is in thermal equilibrium in the plasma, the longitudinal diffusion coefficient $D_{\parallel}$ and the friction coefficient ν are related through Einstein's formula (see Ichimaru, 1973),

$$D_{\parallel}(\mathbf{v}) = -\frac{2T}{m} \nu(v) \tag{1.13.11}$$

and the isotropic stationary solution of the Fokker-Planck equation (1.13.9) is given by,

$$f(v)d\mathbf{v} = A\exp\left[\int \frac{2\nu v}{D_{\parallel}} dv\right] 4\pi v^2 dv$$

$$= A\exp(-\frac{mv^2}{2T})4\pi v^2 dv, \qquad (1.13.12)$$

which is a Maxwellian.

The discussions presented so far have several interesting implications. First, we have shown that the Maxwell distribution is a result of maximizing the entropy. This result can be obtained without using specific information about the collision process which leads to the distribution. We then showed that a collisional process (e.g., Coulomb collision) which satisfies Einstein's formula relating friction and the diffusion coefficients can give rise to the Maxwell distribution as the stationary isotropic solution of the Fokker-Planck equation. These results indicate that if Einstein's formula is valid, then Maxwell's distribution results. This result is consistent with the fact that Einstein's relation is derived for a system of thermal equilibrium.

In space plasmas, however, the observed particle distribution function often deviates from a Maxwellian distribution. This is a consequence of the fact that the Coulomb mean free path is very large and the plasma does not have sufficient time to equilibrate. In particular if the Coulomb field is induced by nonequilibrium electromagnetic radiation (in addition to the electrostatic field of the individual particles), the Einstein relation will break down. Hasegawa et al. (1985) showed that electromagnetic radiation modifies the diffusion coefficient to

$$D_{\parallel} = D_{\parallel}^{c} + D_{\parallel}^{R}$$

$$= D_{\parallel}^{c}(1+\alpha v^2), \qquad (1.13.13)$$

where $D_{\parallel}^{c}$ and $D_{\parallel}^{R}$ are the diffusion coefficients produced by the equilibrium Coulomb field and the radiation induced field respectively and α is a constant which is proportional to the radiation intensity. The isotropic stationary solution of the Fokker-Planck equation will then become, from Eqs. (1.13.11) and (1.13.12),

$$f(v)d\mathbf{v} = A\exp\left[\int \frac{2\nu v}{D_{\parallel}^{c}(1+\alpha v^2)} dv\right] 4\pi v^2 dv$$

$$= A\left[1+\frac{v^2}{2\kappa v_T^2}\right]^{-\kappa} 4\pi v^2 dv, \qquad (1.13.14)$$

where κ is a constant which is inversely proportional to the radiation intensity, v_T is the thermal speed and the normalization constant A is given by

$$A = \frac{2\kappa - 3}{4\sqrt{2}(\pi\kappa)^{3/2} v_T^2} \frac{\Gamma(\kappa)}{\Gamma(\kappa - 1/2)}, \tag{1.13.15}$$

and Γ is the Γ function.

We note that the distribution function Eq. (1.13.14) becomes a Maxwellian at the limit of $\kappa \to \infty$ or in the absence of the superthermal electromagnetic radiation.

A unique feature of this distribution function, which is sometimes called the κ–distribution function, is the fact that in a large velocity region, it gives a power-law distribution in energy E, such that $f \sim (E/E_o)^{-\kappa}$; the transition energy E_o is related to the thermal energy $E_T = mv_T^2$ through

$$\frac{E_o}{E_T} = \kappa. \tag{1.13.16}$$

Particle distributions observed in space can often be fit very well with this κ–distribution function.

References for Chapter 1

H. Goldstein, *Classical Mechanics*, Addison-Wesley, Reading, MA., 1950.

A. Hasegawa, *Plasma Instabilities and Nonlinear Effects*, Springer-Verlag, Berlin, Heidelberg, New York, 1975.

A Hasegawa, K. Mima and M. Dwong-van, *Phys. Rev. Lett. 54*, 2608, 1985.

S. Ichimaru, *Basic Principles of Plasma Physics*, The Benjamin Pub. Co. London, 1973.

Yu. L. Klimontovich, *The Statistical Theory of Non-Equilibrium Processes in a Plasma*, MIT Press, Cambridge, MA., 1967.

M. A. Leontovich (Editor), *Review of Plasma Physics*, vol. 1-6, Consultants Bureau, New York, 1965.

C. L. Longmire, *Elementary Plasma Physics*, Interscience Pub., Wiley & Sons, New York, 1963.

M. Schulz and L. J. Lanzerotti, *Particle Diffusion in the Radiation Belts*, Springer-Verlag, Berlin, Heidelberg, New York, 1974.

V. D. Shafranov, Plasma Equilibrium in a Magnetic Field, *Rev. of Plasma Physics*, Ed. by M. A. Leontovich, Consultants Bureau, New York, 1966.

Chapter 2 Small Amplitude Waves

2.1 Introduction

Electromagnetic waves are modified as they propagate in a plasma because the electrons and ions which are moved by the electromagnetic field of the wave produce charge and current density variations which in turn contribute to the source of the electromagnetic field. Such modification occurs when the frequency of the wave, ω, is comparable to or lower than the characteristic frequencies of the movement of plasma particles. In most plasmas, the highest characteristic frequencies are the electron plasma frequency, ω_{pe}, and the cyclotron frequency, ω_{ce}, (see Sect. 1.4), while the lowest frequencies are those associated with the "fluid" properties of the plasma as described by MHD equations.

Small amplitude waves, which will be considered in this chapter, are characterized by a linear response of the plasma and thus may be treated as a Fourier mode with the amplitude A_k and the phase $(\mathbf{k}\cdot\mathbf{r}-\omega t)$, i.e.,

$$A(\mathbf{r},t)=A_k e^{i(\mathbf{k}\cdot\mathbf{r}-\omega t)}+\text{c.c.}, \tag{2.1.1}$$

where c.c. indicates the complex conjugate.

The vector $\mathbf{k}$ is called the wave vector and its direction designates the direction of the wave propagation; its magnitude $k=\sqrt{\mathbf{k}\cdot\mathbf{k}}$ is called the wave number and is related to the wavelength λ through

$$k=\frac{2\pi}{\lambda}. \tag{2.1.2}$$

$\omega(\mathbf{k})$ is the angular frequency of the wave and in general in a homogeneous plasma is a function of the wave vector $\mathbf{k}$. ω is related to the wave frequency f and period T through

$$\omega=2\pi f=\frac{2\pi}{T}. \tag{2.1.3}$$

The nature of a small amplitude wave can be characterized by the wave vector dependence of the wave frequency $\omega(\mathbf{k})$ which is called the dispersion relation. Given the dispersion relation, the phase velocity $\mathbf{v}_{ph}$ of the wave is defined as

$$\mathbf{v}_{\mathrm{ph}} = \frac{\omega}{\mathbf{k}} = \frac{\omega}{\mathrm{k}}\,\hat{\mathbf{k}} \tag{2.1.4}$$

and the group velocity $\mathbf{v}_g$ of the wave is defined as

$$\mathbf{v}_g = \frac{\partial \omega}{\partial \mathbf{k}} = \frac{\partial \omega}{\partial \mathrm{k}}\,\hat{\mathbf{k}} + \frac{1}{\mathrm{k}}\frac{\partial \omega}{\partial \theta}\hat{\boldsymbol{\theta}}, \tag{2.1.5}$$

where $\hat{\mathbf{k}}$ is the unit vector in the direction of $\mathbf{k}$ and $\hat{\boldsymbol{\theta}}$ is the unit vector normal to $\mathbf{k}$ and coplanar with $\mathbf{v}_g$ and $\mathbf{k}$.

The wave amplitude $A_{\mathbf{k}}$ is a complex function of ω and $\mathbf{k}$ and is obtained either by the Fourier expansion of $A(\mathbf{r}, t)$ (for the periodic case) or by a Fourier transform of $A(\mathbf{r}, t)$ (for the nonperiodic case).

Waves in a plasma may be categorized in terms of frequency range, or in terms of physical properties such as the direction of polarization. However, here we will emphasize the physical nature of waves under various plasma conditions rather than categorization. For this purpose we first consider the simplest situation, waves in a cold unmagnetized plasma (Sect. 2.2); we will then extend the result to waves in a cold but magnetized plasma (Sect. 2.3).

Near the resonance where $\mathrm{k} \gg \omega/c$, a wave tends to have an electrostatic nature. If we assume $\mathbf{E} = -\nabla\phi$, the dispersion relation in the frequency range near the resonance simplifies significantly. The mode derived in this way is called quasi-electrostatic and its dispersion relation is introduced in Section 2.4.

With experience in the treatment of small amplitude waves and the derivation of a dispersion relation, in Section 2.5 we will consider the magnetohydrodynamic waves in a plasma described by MHD equations. In addition, in this section waves which appear in an MHD plasma are compared with those derived in Section 2.2 and the relations between them will be discussed. It will be shown that the MHD waves bring about a new wave associated with a pressure perturbation which is absent in cold plasmas.

In order to describe the wave characteristics associated with the pressure perturbation, we derive the wave dispersion relation using the Vlasov equation and show the existence of wave-particle interactions and associated (Landau) damping of the pressure-driven wave in Section 2.5. In addition the Vlasov plasma is shown to produce a new class of waves associated with the Larmor motion of the individual particles (the Bernstein wave) at the harmonics of the cyclotron frequencies of electrons and ions. These waves are discussed in Section 2.6.

While Sections 2.2 to 2.6 treat waves in a homogeneous plasma, the remainder of the chapter discusses waves in an inhomogeneous plasma. Inhomogeneity brings in a new set of waves which propagate along the surface of the inhomogeneity, called surface waves (Sect. 2.7). In addition, the surface wave faces resonant absorption in a collisionless plasma analogous to Landau damping (Sect. 2.8). In addition to these nonlocal or global waves, the inhomogeneous pressure produces a local wave called a drift wave (Sect. 2.9), and a local inhomogeneous magnetic field produces a change in the dispersion relation of the MHD waves (Sect. 2.10).

A plasma with inhomogeneous pressure has free energy associated with the expansion force. Hence, a wave in such a plasma can become unstable even if the plasma satisfies an equilibrium condition. In Sections 2.9 and 2.10, a brief introduction to the instability conditions of local waves is given as an introduction to Volume II where the instability problems are treated exclusively. In the last two Sections 2.11 and 2.12, the quasiparticle picture of the wave is presented by introducing the concepts of wave energy and momentum densities (Sect. 2.11) and the wave kinetic equation.

2.2 Waves in a Cold Plasma Without a Magnetic Field

A cold plasma is a strange terminology because a plasma is a hot material by definition. A plasma is assumed to be "cold", when the thermal effects are negligible. In a wave analysis, it usually means that the phase velocity of the wave is much larger than the thermal velocity of the plasma particles. Then the collective effect due to the thermal motion of the plasma particles which produces the sound wave and the discrete effect of the individual particle motions which produces the wave-particle resonant interactions are negligible.

Small amplitude waves in a homogeneous plasma may be described by the Fourier transformed Maxwell equations,

$$\mathbf{k} \times \mathbf{E}_{\mathbf{k}} = \omega \mathbf{B}_{\mathbf{k}} \tag{2.2.1}$$

$$\mathbf{k} \times \mathbf{B}_{\mathbf{k}} = -\frac{\omega}{c^2}\left(\mathbf{E}_{\mathbf{k}} - \frac{\mathbf{J}_{\mathbf{k}}}{i\omega\epsilon_0}\right), \tag{2.2.2}$$

where $\mathbf{J}_{\mathbf{k}}$ is the Fourier transform of the current density field induced in the plasma due to the motion of plasma particles in response to the electromagnetic field $\mathbf{E}_{\mathbf{k}}$ and $\mathbf{B}_{\mathbf{k}}$. If we substitute Eq. (2.2.1) into (2.2.2), we have

$$\mathbf{k}\times(\mathbf{k}\times\mathbf{E_k})+\frac{\omega^2}{c^2}(\mathbf{E_k}-\frac{\mathbf{J_k}}{i\omega\epsilon_o})=0. \qquad (2.2.3)$$

Equation (2.2.3) shows that if $\mathbf{J_k}$ is expressed in terms of $\mathbf{E_k}$, we can derive the dispersion relation $\omega=\omega(\mathbf{k})$ by eliminating $\mathbf{E_k}$ from the resultant equation. Hence a wave in a homogeneous plasma can be found by obtaining the linear response in the current density by appropriate kinetic equations.

If we consider a plasma without an external magnetic field, the linearized equation of motion for cold electron and ion fluids can be obtained from Eq. (1.7.4) by setting the magnetic field to zero. Thus,

$$-i\omega m_e \mathbf{v}_{e\mathbf{k}}=-e\mathbf{E_k} \qquad (2.2.4)$$

$$-i\omega m_i \mathbf{v}_{i\mathbf{k}}=e\mathbf{E_k}, \qquad (2.2.5)$$

while the current density is given by

$$\mathbf{J_k}=en_o(\mathbf{v}_{i\mathbf{k}}-\mathbf{v}_{e\mathbf{k}}). \qquad (2.2.6)$$

If we substitute Eqs. (2.2.4) and (2.2.5) into (2.2.6), we can express $\mathbf{J_k}$ in terms of $\mathbf{E_k}$,

$$\mathbf{J_k}=-i\omega\epsilon_o(-\frac{\omega_p^2}{\omega^2})\mathbf{E_k}, \qquad (2.2.7)$$

where

$$\omega_p=\sqrt{\omega_{pi}^2+\omega_{pe}^2}\simeq\omega_{pe}$$

is the plasma frequency. Equation (2.2.7) shows the current density induced in the unmagnetized cold plasma by the wave electric field $\mathbf{E_k}$. If we combine the induced current of Eq. (2.2.7) with the displacement current $\mathbf{J}_d=-i\omega\epsilon_o\mathbf{E_k}$, the total current $\mathbf{J}_\mathbf{k}^t$ is

$$\mathbf{J}_\mathbf{k}^t=\mathbf{J}_d+\mathbf{J_k}$$

$$\equiv-i\omega\epsilon_o\epsilon(\omega)\mathbf{E_k}, \qquad (2.2.8)$$

where the dielectric constant ϵ of the cold and unmagnetized plasma is given by

$$\epsilon(\omega)=1-\frac{\omega_p^2}{\omega^2}. \qquad (2.2.9)$$

The Maxwell equation (2.2.3) when combined with the induced current of Eq. (2.2.7) reads

$$\mathbf{k}\times(\mathbf{k}\times\mathbf{E}_k)+\frac{\omega^2}{c^2}(1-\frac{\omega_p^2}{\omega^2})\mathbf{E}_k=0. \qquad (2.2.10)$$

We see that two types of waves exist depending on the polarization. One is a transverse wave which has its electric field vector perpendicular to the wave vector **k**, i.e., $\mathbf{k}\cdot\mathbf{E}_k=0$. For this wave $\mathbf{k}\times(\mathbf{k}\times\mathbf{E}_k)=-k^2\mathbf{E}_k$; thus the dispersion relation is given by

$$k^2=\frac{\omega^2}{c^2}(1-\frac{\omega_p^2}{\omega^2}). \qquad (2.2.11)$$

This dispersion relation shows that an electromagnetic wave in a vacuum whose dispersion relation is $k^2=\omega^2/c^2$, is modified by the presence of the plasma. In particular for $\omega<\omega_p$, k^2 becomes negative, indicating that an electromagnetic wave cannot propagate in this frequency range (see Fig. 2.1).

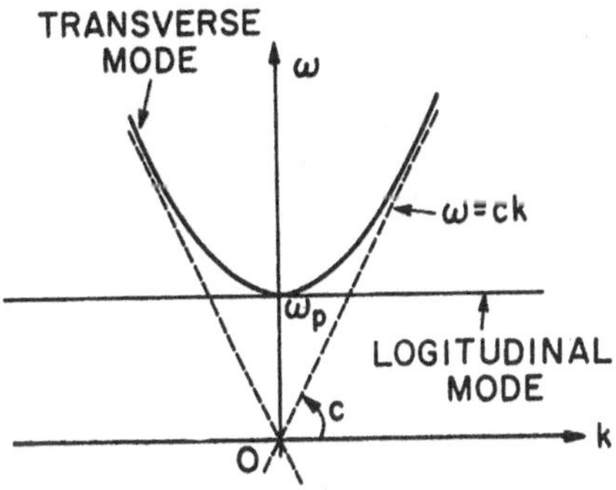

Fig. 2.1. Dispersion relation of electromagnetic waves in a cold, unmagnetized plasma

The other type of wave is a longitudinal wave whose electric field is polarized in the direction of the wave vector **k**. For this wave $\mathbf{k}\times\mathbf{E}_k=0$ and from Eq. (2.2.10) the dispersion relation is given by

$$\omega^2=\omega_p^2. \qquad (2.2.12)$$

This mode represents the longitudinal plasma oscillation discussed in Section 1.4.

2.3 Waves in a Cold Plasma with a Uniform Magnetic Field

Let us now consider the effect of an externally applied uniform magnetic field $\mathbf{B}_o$ on wave propagation in a cold plasma. The equation of motion for the ion fluid from Eq. (1.7.8) reads,

$$-i\omega \mathbf{v}_{ik} = \frac{e}{m_i}\mathbf{E}_k + \mathbf{v}_{ik}\times\omega_{ci}\,, \tag{2.3.1}$$

where $\omega_{ci} = e\mathbf{B}_o/m_i$ is the vector ion cyclotron frequency directed parallel to $\mathbf{B}_o$. Taking the cross-product of ω_{ci} with Eq. (2.3.1) and substituting the result back into Eq. (2.3.1), we have

$$\mathbf{v}_{ik} = \frac{e/m_i}{\omega_{ci}^2-\omega^2}[\mathbf{E}_k\times\omega_{ci} + \frac{i}{\omega}\omega_{ci}(\mathbf{E}_k\cdot\omega_{ci}) - i\omega\mathbf{E}_k]\,. \tag{2.3.2}$$

This expression shows that the velocity field has components in the direction perpendicular to $\mathbf{E}_k$ and $\mathbf{B}_o$, that parallel to $\mathbf{B}_o$ and that parallel to $\mathbf{E}_k$. The first term and the perpendicular component of the last term correspond to the $\mathbf{E}\times\mathbf{B}$ and the polarization drifts respectively, at the zero frequency limit; the second term combined with the parallel component of the last term represents the plasma oscillation.

We can derive a similar expression for the electron velocity field, from which the current density is obtained using Eq. (2.2.6).

If we evaluate the total current density using Eq. (2.2.8), we now see that the plasma dielectric constant becomes a tensor,

$$\mathbf{J}_k^t = -i\omega\epsilon_o \overleftrightarrow{\epsilon}\cdot\mathbf{E}_k\,. \tag{2.3.3}$$

If we employ a right handed cartesian coordinate system with the z axis in the direction of the magnetic field $\mathbf{B}_o$, then each component of the dielectric tensor becomes,

$$\epsilon_{xx} = \epsilon_{yy} = \frac{1}{2}(R+L) = 1 - \sum_j \frac{\omega_{pj}^2}{\omega^2-\omega_{cj}^2} \tag{2.3.4}$$

$$\epsilon_{xy} = -\epsilon_{yx} = -\frac{i}{2}(R-L) = i\sum_j \frac{\omega_{pj}^2\omega_{cj}}{\omega(\omega^2-\omega_{cj}^2)} \tag{2.3.5}$$

$$\epsilon_{xz} = \epsilon_{yz} = \epsilon_{zx} = \epsilon_{zy} = 0 \tag{2.3.6}$$

$$\epsilon_{zz} = 1 - \sum_j \frac{\omega_{pj}^2}{\omega^2}, \tag{2.3.7}$$

where

$$R = 1 - \sum_j \frac{\omega_{pj}^2}{\omega^2} \left(\frac{\omega}{\omega + \omega_{cj}} \right) \tag{2.3.8}$$

$$L = 1 - \sum_j \frac{\omega_{pj}^2}{\omega^2} \left(\frac{\omega}{\omega - \omega_{cj}} \right). \tag{2.3.9}$$

Summations in Eqs. (2.3.7) to (2.3.9) are over electrons and ions, $\omega_{cj} = \omega_{ci}$ for ions, $\omega_{cj} = -\omega_{ce}$ for electrons while $\omega_{pj} = \omega_{pi}$ for ions and ω_{pe} for electrons.

Maxwell's equation now reads, with Eqs. (2.2.3) and (2.3.3),

$$\mathbf{n} \times (\mathbf{n} \times \mathbf{E}_k) + \overleftrightarrow{\epsilon} \cdot \mathbf{E}_k = 0, \tag{2.3.10}$$

where **n** is the vector index of refraction,

$$\mathbf{n} = \frac{c\mathbf{k}}{\omega}. \tag{2.3.11}$$

Without any loss of generality, we take the direction of wave propagation to be in the y-z plane and introduce the angle θ to indicate the direction between **k** and $\mathbf{B}_o$, as shown in Fig. 2.2; **n** then becomes,

$$\mathbf{n} = (0,\ n \sin\theta,\ \ n \cos\theta),$$

and Eq. (2.3.10) is represented

$$\begin{pmatrix} \epsilon_{xx} - n^2 & \epsilon_{xy} & 0 \\ \epsilon_{yx} & \epsilon_{yy} - n^2\cos^2\theta & n^2\cos\theta\sin\theta \\ 0 & n^2\cos\theta\sin\theta & \epsilon_{zz} - n^2\sin^2\theta \end{pmatrix} \begin{pmatrix} E_x \\ E_y \\ E_z \end{pmatrix} = 0. \tag{2.3.12}$$

The dispersion relation for a wave in a cold, magnetized plasma is obtained by setting the determinant of the matrix in this expression to zero. The result can be put in the following form (Stix, 1962),

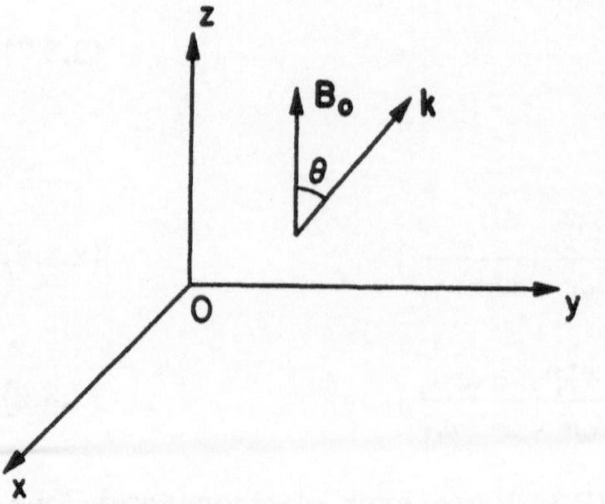

Fig. 2.2. Direction of the wave vector **k** with respect to the magnetic field $\mathbf{B}_0$

$$\tan^2\theta = \frac{-\epsilon_{zz}(n^2-R)(n^2-L)}{(\epsilon_{xx}n^2-RL)(n^2-\epsilon_{zz})}\,. \tag{2.3.13}$$

The dispersion relation thus depends on the angle of propagation with respect to the external magnetic field. For example, if we take $\theta = 0$, i.e., $\mathbf{k} \| \mathbf{B}_0$, the dispersion relation is given from Eq. (2.3.13),

$$\epsilon_{zz} = 0 \tag{2.3.14a}$$

$$n^2 = R \tag{2.3.15a}$$

and

$$n^2 = L\,. \tag{2.3.16a}$$

Equation (2.3.14a) gives the longitudinal plasma oscillation involving E_z,

$$\omega^2 = \omega_p^2 \tag{2.3.14b}$$

which is identical to the longitudinal mode in the absence of the magnetic field, Eq. (2.2.12).

Equation (2.3.15a) gives a transverse mode involving E_x and E_y

$$\frac{c^2k^2}{\omega^2} = 1 - \frac{\omega_{pi}^2}{\omega^2}\frac{\omega}{\omega+\omega_{ci}} - \frac{\omega_{pe}^2}{\omega^2}\frac{\omega}{\omega-\omega_{ce}}\,. \tag{2.3.15b}$$

This mode has two branches, the lower frequency branch has a resonance ($k \rightarrow \infty$) at the electron cyclotron frequency ($\omega = \omega_{ce}$). Because of the resonance at the electron cyclotron frequency, this wave is called the electron cyclotron wave.

At intermediate frequencies, $\omega_{ci} << \omega << \omega_{ce}$, the electron cyclotron wave has a dispersion relation given by

$$\omega \simeq k^2 c^2 \frac{\omega_{ce}}{\omega_{pe}^2}. \tag{2.3.17a}$$

The group velocity is given by

$$\frac{\partial \omega}{\partial k} = 2c \frac{\sqrt{\omega \omega_{ce}}}{\omega_{pe}}, \tag{2.3.17b}$$

which is proportional to $\sqrt{\omega}$. This wave is often observed on the ground after propagating through the magnetosphere from the other hemisphere where it was excited by lightning. Because the higher frequency component of the wave energy arrives earlier, the wave as an audio signal sounds like a descending whistling tone, and is often called a whistler wave.

The higher frequency branch has a cut off frequency at $\omega = (\omega_{ci} + \sqrt{\omega_{ce}^2 + 4\omega_{pe}^2})/2$ and becomes the electromagnetic wave in vacuum, $\omega = ck$, at a large frequency. In its entire frequency range, the wave is right hand polarized, because, from Eqs. (2.3.12) and (2.3.15a)

$$\frac{E_y}{E_x} = \frac{n^2 - \epsilon_{xx}}{\epsilon_{xy}} = \frac{R - \epsilon_{xx}}{\epsilon_{xy}} = i. \tag{2.3.18}$$

Hence with $E_x = \text{Re}\, e^{-i\omega t} = \cos\omega t$, $E_y = \text{Re}\, ie^{-i\omega t} = \sin\omega t$, the electric field vector is seen to rotate to the right looking toward the z direction, the direction of wave propagation.

At very low frequencies, $\omega << \omega_{ci}$, the dispersion relation reduces to

$$\frac{c^2 k^2}{\omega^2} = 1 + \frac{\omega_{pi}^2}{\omega_{ci}^2} \simeq \frac{\omega_{pi}^2}{\omega_{ci}^2}$$

or

$$\omega^2 = k^2 v_A^2, \tag{2.3.19}$$

where

$$v_A = c\frac{\omega_{ci}}{\omega_{pi}} = \frac{B_0}{\sqrt{m_i n_0 \mu_0}} \tag{2.3.20}$$

is the speed of the Alfveń wave (to be discussed in Sect. 2.5).

The dispersion relation given by Eq. (2.3.16a) gives another transverse wave also involving E_x and E_y,

$$\frac{c^2k^2}{\omega^2}=1-\frac{\omega_{pi}^2}{\omega^2}\frac{\omega}{\omega-\omega_{ci}}-\frac{\omega_{pe}^2}{\omega^2}\frac{\omega}{\omega+\omega_{ce}}. \tag{2.3.16b}$$

This wave is left hand polarized, has a resonance at $\omega=\omega_{ci}$, and a cut off at $\omega=(-\omega_{ce}+\sqrt{\omega_{ce}^2+4\omega_{pe}^2})/2$. The lower frequency branch is called the ion cyclotron wave. If we replace ω by $-\omega$, this dispersion relation is identical to that of the electron cyclotron wave. Hence the dispersion relation of both of these waves may be plotted in one figure by including the $\omega<0$ region as in Fig. 2.3.

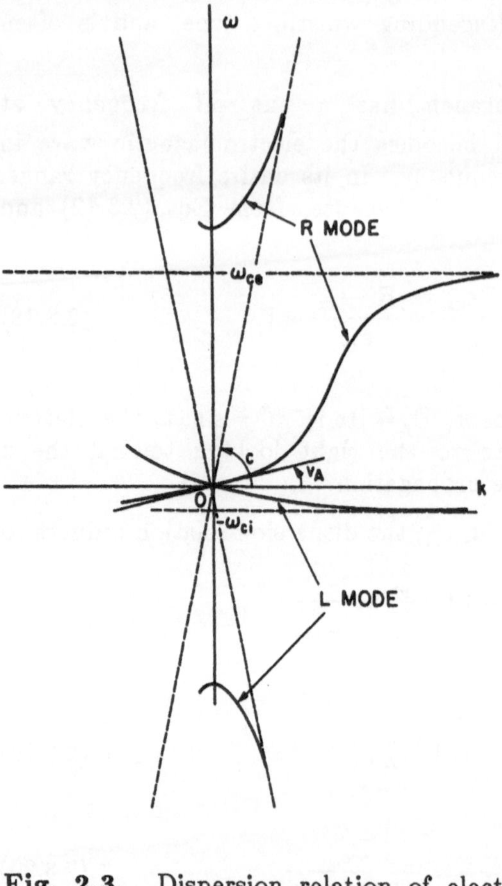

Fig. 2.3. Dispersion relation of electron cyclotron ($\omega>0$) and ion cyclotron ($\omega<0$) waves

As in the case of the electron cyclotron wave, at the zero frequency limit the ion cyclotron wave also has the Alfveń speed, i.e., $\omega = kv_A$.

For perpendicular propagation, $\theta = \pi/2$ in Eq. (2.3.13) gives

$$n^2 = RL/\epsilon_{xx} \tag{2.3.21a}$$

and

$$n^2 = \epsilon_{zz} . \tag{2.3.22a}$$

In this case the corresponding wave vector is in the y-direction. The electric field associated with the mode given in Eq. (2.3.21a) has components in both the x and y directions; hence the wave is a mixed transverse and longitudinal wave. The dispersion relation (2.3.21a) may be cast into the form,

$$\frac{c^2k^2}{\omega^2} = \frac{(\omega^2-\omega_R^2)(\omega^2-\omega_L^2)}{(\omega^2-\omega_{LH}^2)(\omega^2-\omega_{UH}^2)} . \tag{2.3.21b}$$

Here ω_R and ω_L satisfy

$$\omega^2 - \omega\omega_{ce} - \omega_{ce}\omega_{ci} - \omega_{pe}^2 = 0 \tag{2.3.23}$$

and

$$\omega^2 + \omega\omega_{ce} - \omega_{ce}\omega_{ci} - \omega_{pe}^2 = 0 \tag{2.3.24}$$

while ω_{LH} is the lower hybrid resonant frequency satisfying

$$\frac{1}{\omega_{LH}^2} = \frac{1}{\omega_{ci}^2 + \omega_{pi}^2} + \frac{1}{\omega_{ci}\omega_{ce}} , \tag{2.3.25a}$$

or

$$\omega_{LH}^2 \simeq \omega_{pi}^2 \frac{\omega_{ce}^2}{\omega_{ce}^2 + \omega_{pe}^2} \tag{2.3.25b}$$

and ω_{UH} is the upper hybrid resonant frequency given by

$$\omega_{UH}^2 = \omega_{pe}^2 + \omega_{ce}^2 . \tag{2.3.26}$$

The wave whose dispersion relation is given by Eq. (2.3.21) is called the extraordinary wave. It has two resonances and two cut off frequencies and the dispersion relation is shown in Fig. 2.4. At $\omega << \omega_{ci}$, the dispersion relation is given by $\omega \simeq kv_A$, and is called the compressional Alfveń wave because $\mathbf{k} \cdot \mathbf{v}_k \neq 0$ for this mode.

The dispersion relation given by Eq. (2.3.22a) has the form,

$$\omega^2 = k^2 c^2 + \omega_{pe}^2 \tag{2.3.22b}$$

and is called the ordinary wave. It has a structure identical to the electromagnetic wave in the absence of the magnetic field given by Eq. (2.2.11). The electric field of the ordinary wave is polarized in the direction of the external magnetic field, hence the induced motions of plasma particles are not influenced by the magnetic field. This fact leads to the dispersion relation which is not affected by the magnetic field.

2.4 Quasi-Electrostatic Waves

The lower and upper hybrid waves introduced in Section 2.3 become almost purely longitudinal near the resonance, that is, the y component of electric field vector becomes dominant. In general, at a frequency close to the resonance, where $k \rightarrow \infty$, the wave phase velocity becomes much smaller than the speed of light, hence the wave characteristic approaches an electrostatic mode. We call these modes quasi-electrostatic modes. The dispersion relation of a quasi-electrostatic mode can be obtained by using the electrostatic potential ϕ, $\mathbf{E} = -\nabla\phi$ or $\mathbf{E}_k = -i\mathbf{k}\phi_k$, substituting into the Maxwell equation

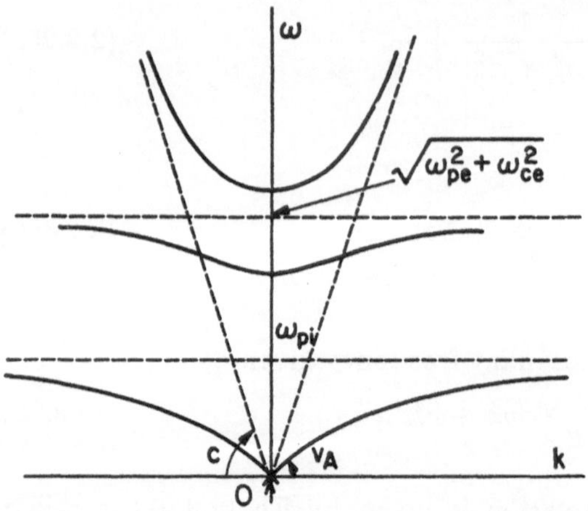

Fig. 2.4. Dispersion relation of extraordinary waves

(2.3.10) with the dielectric tensor ϵ and taking the divergence of both sides,

$$\mathbf{k}\cdot(\overleftrightarrow{\epsilon}\cdot\mathbf{k})=0. \tag{2.4.1}$$

For a cold plasma in an external magnetic field, Eq. (2.4.1) may be expressed explicitly using the perpendicular, $k_\perp$, and parallel, $k_\parallel$, components of the wave vectors

$$k_\perp^2\epsilon_\perp+k_\parallel^2\epsilon_\parallel=0, \tag{2.4.2a}$$

where $k_\perp^2=k_x^2+k_y^2$, $\epsilon_\perp=\epsilon_{xx}=\epsilon_{yy}$ and $\epsilon_\parallel=\epsilon_{zz}$, or using Eqs. (2.3.4) and (2.3.7)

$$k_\perp^2\left(1-\frac{\omega_{pe}^2}{\omega^2-\omega_{ce}^2}-\frac{\omega_{pi}^2}{\omega^2-\omega_{ci}^2}\right)$$

$$+k_\parallel{}^2\left(1-\frac{\omega_{pe}^2}{\omega^2}-\frac{\omega_{pi}^2}{\omega^2}\right)=0. \tag{2.4.2b}$$

For parallel propagation, $k_\perp=0$, and Eq. (2.4.2) gives

$$\omega^2=\omega_{pe}^2+\omega_{pi}^2$$

which is the longitudinal mode shown in Eq. (2.3.14). For perpendicular propagation, $k_\parallel=0$, and Eq. (2.4.2) gives two roots,

$$\omega^2\simeq\omega_{pe}^2+\omega_{ce}^2=\omega_{UH}^2 \tag{2.4.3}$$

and

$$\omega^2\simeq\omega_{pi}^2\frac{\omega_{ce}^2}{\omega_{ce}^2+\omega_{pe}^2}=\omega_{LH}^2. \tag{2.4.4}$$

2.5 MHD Waves

In order to consider the effect of plasma pressure on wave propagation, let us now consider waves with frequencies much smaller than the ion cyclotron frequency. In this frequency range, the plasma behavior may be described by the magnetohydrodynamic (MHD) equations introduced in Section 1.6.

Here we consider MHD waves in a uniform plasma with density n_0 in a uniform magnetic field $\mathbf{B}_0$. To consider small amplitude waves, we first linearize the MHD equations. Using the subscript 1 for the linearized quantities we have,

$$m_i n_o \frac{\partial \mathbf{v}_1}{\partial t} = \mathbf{J}_1 \times \mathbf{B}_o - \nabla p_1 \tag{2.5.1a}$$

$$\mathbf{E}_1 + \mathbf{v}_1 \times \mathbf{B}_o = 0 \tag{2.5.2a}$$

$$\frac{\partial n_1}{\partial t} + \nabla \cdot (n_o \mathbf{v}_1) = 0 \tag{2.5.3a}$$

$$\nabla \times \mathbf{E}_1 = - \frac{\partial \mathbf{B}_1}{\partial t} \tag{2.5.4a}$$

$$\nabla \times \mathbf{B}_1 = \mu_o \mathbf{J}_1 \tag{2.5.5a}$$

$$\nabla \cdot \mathbf{B}_1 = 0 \tag{2.5.6a}$$

$$\frac{\partial p_1}{\partial t} = \gamma T_o \frac{\partial n_1}{\partial t} \; . \tag{2.5.7a}$$

We then use the Fourier amplitude expressions as given in Eq. (2.1.1) for the linearized quantities such that

$$\mathbf{v}_1(\mathbf{r},t) \equiv \mathbf{v}_\mathbf{k} e^{i(\mathbf{k}\cdot\mathbf{r} - \omega t)} + \text{c.c.}$$

Then Eqs. (2.5.1a) to (2.5.7a) become

$$- i\omega m_i n_o \mathbf{v}_\mathbf{k} = \mathbf{J}_\mathbf{k} \times \mathbf{B}_o - i\mathbf{k} p_\mathbf{k} \tag{2.5.1b}$$

$$\mathbf{E}_\mathbf{k} + \mathbf{v}_\mathbf{k} \times \mathbf{B}_o = 0 \tag{2.5.2b}$$

$$\omega n_\mathbf{k} - n_o \mathbf{k} \cdot \mathbf{v}_\mathbf{k} = 0 \tag{2.5.3b}$$

$$\mathbf{k} \times \mathbf{E}_\mathbf{k} = \omega \mathbf{B}_\mathbf{k} \tag{2.5.4b}$$

$$i\mathbf{k} \times \mathbf{B}_\mathbf{k} = \mu_o \mathbf{J}_\mathbf{k} \tag{2.5.5b}$$

$$\mathbf{k} \cdot \mathbf{B}_\mathbf{k} = 0 \tag{2.5.6b}$$

$$p_\mathbf{k} = \gamma T_o n_\mathbf{k} \; . \tag{2.5.7b}$$

There are two types of waves. One is a torsional wave satisfying the incompressible condition, $\nabla \cdot \mathbf{v}_1 = 0$; the other is a compressional wave, for which $\nabla \cdot \mathbf{v}_1 \neq 0$. The torsional wave accompanies vorticity in the direction of the ambient magnetic field, $\Omega = (\nabla\times\mathbf{v}_1)\cdot\mathbf{B}_o/B_o$. From Eq. (2.5.1b) the vorticity satisfies,

$$-\,\omega m_i n_o \Omega_k = (\mathbf{k}\cdot\mathbf{B}_o)(\mathbf{J}_k\cdot\mathbf{B}_o)\,\frac{1}{B_o}\,. \tag{2.5.8}$$

From Eqs. (2.5.2b), (2.5.4b) we have

$$\mathbf{B}_k = \frac{1}{\omega}\,[(\mathbf{k}\cdot\mathbf{v}_k)\mathbf{B}_o - (\mathbf{k}\cdot\mathbf{B}_o)\mathbf{v}_k]\,. \tag{2.5.9}$$

Substituting Eq. (2.5.9) into (2.5.5b) we have

$$\mathbf{J}_k = \frac{i}{\mu_o\omega}[(\mathbf{k}\cdot\mathbf{v}_k)(\mathbf{k}\times\mathbf{B}_o) - (\mathbf{k}\cdot\mathbf{B}_o)(\mathbf{k}\times\mathbf{v}_k)]\,. \tag{2.5.10}$$

The dispersion relation of the *torsional wave* is obtained by substituting Eq. (2.5.10) into Eq. (2.5.8) and using $\mathbf{k}\cdot\mathbf{v}_k = 0$,

$$[\omega^2 m_i n_o \mu_o - (\mathbf{k}\cdot\mathbf{B}_o)^2]\Omega_k = 0\,. \tag{2.5.11}$$

If we take $k_{\|}$ to be the wave vector in the direction of magnetic field, Eq. (2.5.11) gives the dispersion relation of the Alfveń wave,

$$\omega^2 = k_{\|}^2 v_A^2\,, \tag{2.5.12}$$

$$\text{where}\quad v_A = \frac{B_o}{\sqrt{m_i n_o \mu_o}} \tag{2.3.20}$$

is the Alfveń speed introduced in Section 2.3. From the derivation of the wave equation, we see that the Alfveń wave is a hydromagnetic wave which is associated with the fluid vorticity in the direction of the ambient magnetic field. The wave propagates in the direction of the magnetic field. Since the phase velocity of the Alfveń wave is given by the square root of the ratio of the magnetic tension B_o^2/μ_o and the mass density $m_i n_o$, the wave may be regarded as a tensile wave of the magnetic line of force which has distributed weight density given by $m_i n_o$. It is interesting to recognize that an electromagnetic wave can propagate in a fully conducting fluid at a frequency much lower than the cut-off frequency ω_p of the light wave.

The dispersion relation of the *compressional waves* may be obtained by constructing the scalar products of $i\mathbf{k}$ and $i\mathbf{B}_o$ with the equation of motion Eq. (2.5.1b),

$$\omega m_i n_o(\mathbf{k}\cdot\mathbf{v_k}) = i\mathbf{k}\cdot(\mathbf{J_k}\times\mathbf{B_o}) + k^2 p_k \qquad (2.5.13)$$

and

$$\omega m_i n_o(\mathbf{B_o}\cdot\mathbf{v_k}) = (\mathbf{k}\cdot\mathbf{B_o})p_k \ . \qquad (2.5.14)$$

Substituting Eqs. (2.5.10), (2.5.7b) and (2.5.3b) into these equations and eliminating $(\mathbf{B_o}\cdot\mathbf{v_k})$, we have the following dispersion relation for the compressional waves

$$\omega^4 - \omega^2 k^2(v_A^2+v_s^2) + k^2 k_\parallel^2 v_A^2 v_s^2 = 0 \ , \qquad (2.5.15a)$$

where

$$v_s = \left(\frac{\gamma T_o}{m_i}\right)^{1/2} \qquad (2.5.16)$$

is the collisional sound speed (to be distinguished from the collisionless sound speed introduced later). Since the dispersion relation has order of ω^4, we note that there are two types of compressional waves which are coupled. Solving the dispersion relations for ω^2/k^2, we see

$$\frac{\omega^2}{k^2} = \frac{1}{2}\{(v_A^2+v_s^2) \pm [(v_A^2+v_s^2)^2 - 4v_A^2 v_s^2 k_\parallel^2/k^2]^{1/2}\} . \qquad (2.5.15b)$$

For perpendicular propagation, $k_\parallel = 0$, the two waves have the dispersion relations

$$\frac{\omega^2}{k^2} = v_A^2 + v_s^2 \qquad (2.5.17)$$

$$\text{and} \quad \omega = 0 \ , \qquad (2.5.18)$$

while, for parallel propagation $k = k_\parallel$, these waves become

$$\frac{\omega^2}{k^2} = v_A^2 \qquad (2.5.19)$$

and

$$\frac{\omega^2}{k^2} = v_s^2 \ . \qquad (2.5.20)$$

The wave whose dispersion relation is connected to Eq. (2.5.17) (at any angle of propagation) is called the fast magnetosonic wave while that related to Eq. (2.5.18) is called the slow magnetosonic wave. As will be shown in Section 2.6, the slow wave suffers Landau damping in a collisionless low beta plasma. The fast wave suffers transit time damping (an analog of Landau damping in which the wave particle interaction occurs through the $-\nabla_{\parallel}\mu B$ force) in a high beta plasma.

Thus in contrast to the Alfvén wave, the magnetosonic waves have rather short life times.

2.6 Waves in Vlasov Plasmas

A unique feature of waves in a collisionless plasma is wave-particle resonant interaction. Depending on the particle distribution function the wave particle interaction can lead to a loss or a gain in wave energy. When the distribution function is Maxwellian in velocity space and uniform in coordinate space, a wave particle interaction always leads to a loss of the wave energy.

To illustrate wave particle interactions, let us consider the propagation of a slow wave in the direction of the magnetic field in a low beta plasma. Since the slow wave in a low beta plasma does not accompany a magnetic field perturbation, we assume that the wave field is electrostatic and the wave electric field is given by a gradient of potential ϕ_1. The linearized Vlasov equation for electrons and ions are given respectively from Eq. (1.3.10)

$$\frac{\partial f_1^{(e)}}{\partial t} + v\frac{\partial f_1^{(e)}}{\partial x} + \frac{e}{m_e}\frac{\partial \phi_1}{\partial x}\frac{\partial f_0^{(e)}}{\partial v} = 0 \tag{2.6.1}$$

$$\frac{\partial f_1^{(i)}}{\partial t} + v\frac{\partial f_1^{(i)}}{\partial x} - \frac{e}{m_i}\frac{\partial \phi_1}{\partial x}\frac{\partial f_0^{(i)}}{\partial v} = 0\,, \tag{2.6.2}$$

where x is the coordinate of the wave propagation and v is the velocity in the x direction. The field equation is given by Poisson's equation,

$$\frac{\partial^2 \phi_1}{\partial x^2} = -\frac{en_o}{\epsilon_o}\int_{-\infty}^{\infty}(f_1^{(i)} - f_1^{(e)})dv\,. \tag{2.6.3}$$

We use again the Fourier amplitude expression defined as

$$f_1(x,v,t) = f_k(v)e^{i(kx-\omega t)} + \text{c.c.} \tag{2.6.4}$$

One remark needed here is the fact that causability requires that $\text{Im}\omega > 0$ in this expression so that $f_1 \to 0$ at $t \to -\infty$. If we substitute the Fourier amplitude expression (2.6.4) into Eqs. (2.6.1), (2.6.2) and (2.6.3), and eliminate the field variables, we can obtain the dispersion relations of the electrostatic mode in the following form

$$1 - \frac{\omega_{pe}^2}{k^2} \int_{-\infty}^{\infty} \frac{\partial f_0^{(e)}/\partial v}{v-\omega/k} dv - \frac{\omega_{pi}^2}{k^2} \int_{-\infty}^{\infty} \frac{\partial f_0^{(i)}/\partial v}{v-\omega/k} dv = 0 \,, \qquad (2.6.5)$$

where ω_{pj} is the plasma frequency for ions ($j = i$) and electrons ($j = e$).

The collisionless ion acoustic wave can exist, as will be seen later, only when the electron temperature is much higher than the ion temperature. Thus, for ions, we take a "cold" distribution function represented by the delta function,

$$f_0^{(i)} = \delta(v) \,. \qquad (2.6.6)$$

Then the ion contributions I_i in Eq. (2.6.5) become, after integration by parts,

$$I_i \equiv \int_{-\infty}^{\infty} \frac{\partial f_0^{(i)}/\partial v}{v-\omega/k} dv = \int_{-\infty}^{\infty} \frac{f_0^{(i)}}{(v-\omega/k)^2} dv = \left(\frac{k}{\omega}\right)^2 . \qquad (2.6.7)$$

The integral I_e for the electron contributions may be evaluated using the Dirac integral expression and the requirement of $\mathrm{Im}\,\omega > 0$,

$$\int_{-\infty}^{\infty} \frac{g(x)}{x-(y+i0)} dx = P \int \frac{g(x)}{x-y} dx + i\pi g(y) \,, \qquad (2.6.8)$$

where P is the principal value,

$$I_e \equiv \int_{-\infty}^{\infty} \frac{\partial f_0^{(e)}/\partial v}{v-\omega/k} dv$$

$$= P \int_{-\infty}^{\infty} \frac{\partial f_0^{(e)}/\partial v}{v-\omega/k} dv + i\pi \frac{k}{|k|} \left. \frac{\partial f_0^{(e)}}{\partial v} \right|_{v=\omega/k} . \qquad (2.6.9)$$

For the unperturbed distribution function of electrons, we assume a Maxwellian given by

$$f_0^{(e)} = \frac{1}{(2\pi)^{1/2} v_{Te}} e^{-v^2/2v_{Te}^2} . \qquad (2.6.10)$$

Assuming the wave phase velocity is much smaller than the electron thermal speed v_{Te},

$$I_e \simeq \frac{1}{(2\pi)^{1/2} v_{Te}} \int_{-\infty}^{\infty} \frac{1}{v} \cdot \left(- \frac{v}{v_{Te}^2} e^{-v^2/2v_{Te}^2} \right) dv$$

$$+ i\pi \frac{k}{|k|} \left. \frac{\partial f_0^{(e)}}{\partial v} \right|_{v=\omega/k}$$

$$= - \frac{1}{v_{Te}^2} - \left(\frac{\pi}{2}\right)^{1/2} \frac{\omega}{|k| v_{Te}} \frac{1}{v_{Te}^2} . \qquad (2.6.11)$$

The dispersion relation is obtained from Eqs. (2.6.5), (2.6.7) and (2.6.11),

$$1 - \frac{\omega_{pi}^2}{\omega^2} + \frac{\omega_{pe}^2}{k^2 v_{Te}^2} \left(1 + i \left(\frac{\pi}{2}\right)^{1/2} \frac{\omega}{|k| v_{Te}} \right) = 0 . \qquad (2.6.12)$$

For a low frequency limit, $\omega << \omega_{pi}$, the real part of the dispersion relation is readily solvable

$$\omega = k c_s , \qquad (2.6.13)$$

where c_s is the *collisionless ion sound speed* given by

$$c_s = v_{Te} (m_e/m_i)^{1/2} = (T_e/m_i)^{1/2} . \qquad (2.6.14)$$

The wave represented by the dispersion relation (2.6.13) is called the *collisionless ion acoustic (sound) wave.* We note that the dispersion relation also presents a negative imaginary part of ω which is given by

$$\omega_i = - \left(\frac{\pi}{2} \right)^{1/2} \frac{c_s}{v_{Te}} |k| c_s$$

$$= - \left(\frac{\pi m_e}{2 m_i} \right)^{1/2} |k| c_s . \qquad (2.6.15)$$

The negative imaginary part of ω indicates a damping of the wave. It may seem strange to encounter damping of a wave starting from a dissipationless system. The wave energy is absorbed by resonant streaming electrons with their velocity comparable to the wave phase velocity. In the absence of collisions, the resonant particles preserve the wave energy and information but the collective information is dissipated into a continuum of spectrum of the free streaming electrons. The damping of the wave due to this type of wave-particle resonant interaction mechanism is called Landau damping after the name of the discoverer.

We note that the collisionless ion acoustic wave has a phase velocity given by $(T_e/m_i)^{1/2}$, which is different from the collisional (MHD) acoustic wave given in Eq. (2.5.16). If the ion temperature is comparable to the electron temperature, the above derivation of the collisionless dispersion relations indicates that there will be a strong wave-particle interaction between the wave and ions. This happens because the wave phase velocity is comparable to the ion thermal speed. This produces a large imaginary part in the ion contribution of Eq. (2.6.7) and the resultant dispersion relation produces an imaginary part of ω comparable to the real part. Hence in a collisionless plasma, the ion acoustic wave will not exist if $T_i \simeq T_e$.

In the presence of a magnetic field, the Vlasov equation for the unperturbed velocity distribution function f_o becomes

$$(\mathbf{v}\times\mathbf{B}_o)\cdot\frac{\partial f_o}{\partial \mathbf{v}} = 0\,. \tag{2.6.16}$$

Equation (2.6.16) can be satisfied for any function of $v_\perp$ and $v_\parallel$ as

$$f_o(\mathbf{v}) = f_o(v_\perp, v_\parallel), \tag{2.6.17}$$

where $v_\parallel(=v_z)$ and $v_\perp = (v_x^2+v_y^2)^{1/2}$ are the velocities parallel and perpendicular to the magnetic field, and z is taken to be parallel to $\mathbf{B}_o$.

For electrostatic perturbations, the linearized Vlasov equation for a species with charge q and mass m becomes

$$\frac{\partial f_1}{\partial t} + \mathbf{v}\cdot\frac{\partial f_1}{\partial \mathbf{x}} + \frac{q}{m}(\mathbf{v}\times\mathbf{B}_o)\cdot\frac{\partial f_1}{\partial \mathbf{v}} = \frac{q}{m}\frac{\partial \phi_1}{\partial \mathbf{x}}\cdot\frac{\partial f_o}{\partial \mathbf{v}}\,. \tag{2.6.18}$$

In Eq. (2.6.18) f_1 can be obtained either by integrating along the unperturbed orbit or by solving the differential equation by using cylindrical coordinates in velocity space. We take the former method here. If we use the trajectory decided by the stationary field i.e., $\ddot{x} = q/m(\mathbf{v}\times\mathbf{B}_o)$, the left hand side of Eq. (2.6.18) can be written as a total derivative with respect to time df_1/dt. f_1 can then be integrated formally along such a trajectory

$$f_1(\mathbf{x},\mathbf{v},t) = \int_{-\infty}^{t} dt'\,\frac{q}{m}\,\frac{\partial \phi_1(t')}{\partial \mathbf{x}'}\cdot\left.\frac{\partial f_o}{\partial \mathbf{v}}\right|_{\mathbf{v}=\mathbf{v}'(t')}\,, \tag{2.6.19}$$

where the unperturbed trajectory is given by

$$\left.\begin{aligned} x'(t') - x &= \frac{v_\perp}{\omega_c}\{\sin[\omega_c(t'-t) + \theta] - \sin\theta\} \,, \\ y'(t') - y &= \frac{v_\perp}{\omega_c}\{\cos[\omega_c(t'-t) + \theta] - \cos\theta\} \,, \\ z'(t') - z &= v_\parallel(t'-t) \,. \end{aligned}\right\} \tag{2.6.20}$$

The final position at $t' = t$ is chosen to be $x' = x$, $y' = y$ and $z' = z$, while the corresponding velocities are $v_x(t) = v_\perp \cos\theta$, $v_y(t) = -v_\perp \sin\theta$ and $v_z(t) = v_\parallel$. Note that the sense of rotation is left (right) handed with respect to the direction of the magnetic field for protons (electrons) where $\omega_c = \omega_{ci} = eB_o/m_i > 0$ ($\omega_c = -\omega_{ce} = -eB_o/m_e < 0$) is the cyclotron frequency.

If we consider a perturbation of the form $\exp i(\mathbf{k}\cdot\mathbf{x} - \omega t)$ and take the direction of wave propagation in the x,z plane (without loss of generality) such that

$$\mathbf{k}\cdot\mathbf{x} = k_\perp x + k_\parallel z \,, \tag{2.6.21}$$

Eq. (2.6.19) gives

$$f_1(\mathbf{x},\mathbf{v},t) = \frac{iq}{m}\int_{-\infty}^{t} dt' \left.\left(k_\perp \frac{\partial f_o}{\partial v_x} + k_\parallel \frac{\partial f_o}{\partial v_\parallel}\right)\right|_{\mathbf{v}=\mathbf{v}'(t')} \tag{2.6.22}$$

$$\cdot\, \phi_{\mathbf{k}} \exp i[k_\perp x'(t') + k_\parallel z'(t') - \omega t'] \,.$$

If we substitute x' and z' of Eq. (2.6.20) into the above expression and note that $\partial f_o/\partial v_\parallel|_{v_\parallel = v'_\parallel} = \partial f_o/\partial v_\parallel$ (because $v_\parallel$ does not change), while $\partial f_o/\partial v_x|_{v_x = v'_x} = (\partial f_o/\partial v_\perp)(v_\perp/v_x)|_{v_x = v'_x} = \partial f_o/\partial v_\perp \cos[\omega_c(t'-t) + \theta]$ (because $v_\perp$ is also a constant of motion), we can integrate Eq. (2.6.22). We then obtain the perturbed distribution function f_1 as a function of *independent* variables , $\mathbf{x}, \mathbf{v}$ and t. If we then integrate over the velocity space to obtain the charge density perturbation, $\rho_{\mathbf{k}}$, the result reads

$$\rho_{\mathbf{k}} = \epsilon_o \omega_p^2 \sum_{n=-\infty}^{\infty} \int d\mathbf{v} [J_n(\frac{k_\perp v_\perp}{\omega_c})]^2 \frac{k_\parallel(\partial f_o/\partial v_\parallel) + (n\omega_c/v_\perp)(\partial f_o/\partial v_\perp)}{k_\parallel v_\parallel - (\omega - n\omega_c)} \phi_{\mathbf{k}} \,, \tag{2.6.23}$$

where $d\mathbf{v} = 2\pi v_\perp dv_\perp dv_\parallel$ and J_n is the nth order Bessel function of the first kind. In deriving Eq. (2.6.22) use is made of the identity

$$\exp(iz\sin\theta) = \sum_{n=-\infty}^{\infty} J_n(z)\exp(in\theta).$$

The unperturbed distribution function f_0 may be assumed to be an isotropic Maxwellian,

$$f_0(v) = \left(\frac{1}{\sqrt{2\pi}\,v_T}\right)^3 \exp\left(-\frac{v_\perp^2 + v_\parallel^2}{2v_T^2}\right). \tag{2.6.24}$$

Then Eq. (2.6.22) reduces to

$$\rho_k = -\frac{\epsilon_0 \omega_p^2}{v_T^2} \sum_{n=-\infty}^{\infty} e^{-\lambda} I_n(\lambda)$$

$$\left[1 + \frac{\omega}{\sqrt{2}k_\parallel v_T} Z\left(\frac{\omega - n\omega_c}{\sqrt{2}k_\parallel v_T}\right)\right]\phi_k , \tag{2.6.25}$$

where use is made of the relation

$$\int_0^\infty e^{-a^2x^2} J_n^2(x)x\,dx$$

$$= \frac{1}{2a^2} e^{-b^2/2a^2} I_n\left(\frac{b^2}{2a^2}\right).$$

Here,

$$\lambda = \frac{k_\perp^2 v_T^2}{\omega_c^2}, \tag{2.6.26}$$

I_n is the modified Bessel function of the nth order and Z is the plasma dispersion function defined by

$$Z(\varsigma) = \frac{1}{(\pi)^{1/2}} \int_{-\infty}^{\infty} \frac{e^{-x^2}}{x - \varsigma} dx \text{ for } \operatorname{Im}\varsigma > 0 \tag{2.6.27}$$

= analytic continuation of the above integral for $\operatorname{Im}\varsigma < 0$.

The power-series expansion for a small argument and the asymptotic expansion for a large argument of the Z function are given by

$$Z(\varsigma) \simeq i(\pi)^{1/2} e^{-\varsigma^2} - 2\varsigma(1 - \frac{2\varsigma^2}{3}) \quad \text{for } |\varsigma| << 1 \qquad (2.6.28)$$

$$Z(\varsigma) \simeq i(\pi)^{1/2} \sigma e^{-\varsigma^2} - \frac{1}{\varsigma}(1 + \frac{1}{2\varsigma^2}) \quad \text{for } \quad |\varsigma| >> 1, \qquad (2.6.29)$$

where

$$\sigma = \begin{cases} 0 & \text{Im}\,\varsigma > 0 \\ 1 & \text{Im}\,\varsigma = 0 \\ 2 & \text{Im}\,\varsigma < 0. \end{cases}$$

To appreciate the result of the derived charge density perturbation, let us consider an example of an electrostatic wave at frequencies much higher than the ion plasma frequency. Then the dominant charge density perturbation is that of electrons and the dispersion relation is obtained by substituting Eq. (2.6.25) for electrons into Poisson's equation,

$$1 + \frac{\omega_{pe}^2}{k^2 v_{Te}^2} \sum_{n=-\infty}^{\infty} e^{-\lambda_e} I_n(\lambda_e) \left[1 + \frac{\omega}{\sqrt{2} k_{\|} v_{Te}} Z(\frac{\omega - n\omega_{ce}}{\sqrt{2} k_{\|} v_{Te}}) \right] = 0, \qquad (2.6.30)$$

where

$$\lambda_e = \frac{k_\perp^2 v_{Te}^2}{\omega_{ce}^2}. \qquad (2.6.31)$$

Now let us consider an example of a parallel propagating wave, $k_\perp = 0$. In this case since $\lambda_e = 0$, only the $n = 0$ term will contributes (since $I_n(0) = 0$, $n = 1, 2, ..$, while $I_0(0) = 1$). Eq. (2.6.30) then reduces to

$$1 + \frac{\omega_{pe}^2}{k_{\|}^2 v_{Te}^2} \left[1 + \frac{\omega}{\sqrt{2} k_{\|} v_{T_e}} Z\left(\frac{\omega}{\sqrt{2} k_{\|} v_{T_e}} \right) \right] = 0. \qquad (2.6.32)$$

If we further take a small $|k_{\|}|$ limit, from Eq. (2.6.29), $Z(\varsigma) \simeq -1/\varsigma(1 + 1/2\varsigma^2)$, thus Eq. (2.6.32) gives

$$1-\frac{\omega_p^2}{\omega^2}=0, \tag{2.6.33}$$

which gives the dispersion relation of the electron plasma oscillation. Equation (2.6.32) shows, however, that if $\omega/|k_\parallel|v_{Te}$ is of the order of unity, the Z-function will have an appreciably large imaginary part and the plasma oscillation will be heavily damped due to the Landau damping. Since $\omega \simeq \omega_p$, the condition of $\omega/|k_\parallel|v_{Te} \simeq 0(1)$ means that the wave lengths are comparable to the Debye length v_{Te}/ω_p. Since the electrostatic field is shielded beyond the Debye length, the absence of a plasma oscillation for wavelengths shorter than v_{Te}/ω_p is a natural consequence of the Debye shielding effect.

We now look at the perpendicular propagation. In this case, letting $k_\parallel \rightarrow 0$, Eq. (2.6.30) gives

$$1+\frac{\omega_{pe}^2}{k^2 v_{Te}^2}\left[1-\sum_{n=-\infty}^{\infty} e^{-\lambda_e} I_n(\lambda_e)\frac{\omega}{\omega-n\omega_{ce}}\right]=0,$$

which can be further reduced using the relation $\sum_{n=-\infty}^{\infty} I_n(\lambda)e^{-\lambda}=1$, to

$$1-2\frac{\omega_{pe}^2}{\omega_{ce}^2}\sum_{n=1}^{\infty} I_n(\lambda_e)\frac{e^{-\lambda_e}}{\lambda_e}\frac{n^2}{(\omega/\omega_{ce})^2-n^2}=0. \tag{2.6.34}$$

We note that since

$$I_n(z)=\left(\frac{z}{2}\right)^n \sum_{m=0}^{\infty}\frac{(z/2)^{2m}}{m!(m+n)!}$$

$$\text{and } I_n(z) \sim \frac{e^z}{\sqrt{2\pi z}} \quad (z \rightarrow \infty),$$

the resonance $(k \rightarrow \infty)$ occurs at a harmonic of the cyclotron frequency, $n\omega_{ce}$. In addition the cut off $(k=0)$ occurs also at a harmonic of the cyclotron frequency except at $n=1$, where the cut off frequency is given by the upper hybrid frequency, $(\omega_{pe}^2+\omega_{ce}^2)^{1/2}$. The wave given by the dispersion relation, Eq. (2.6.34) is called the Bernstein wave after the discoverer of the wave. The dispersion relation is plotted in Fig. 2.5.

When this dispersion relation is compared with the electrostatic mode in a cold plasma, Eq. (2.4.3), we notice a considerable complexity in waves in a Vlasov plasma. In particular, the structure at the

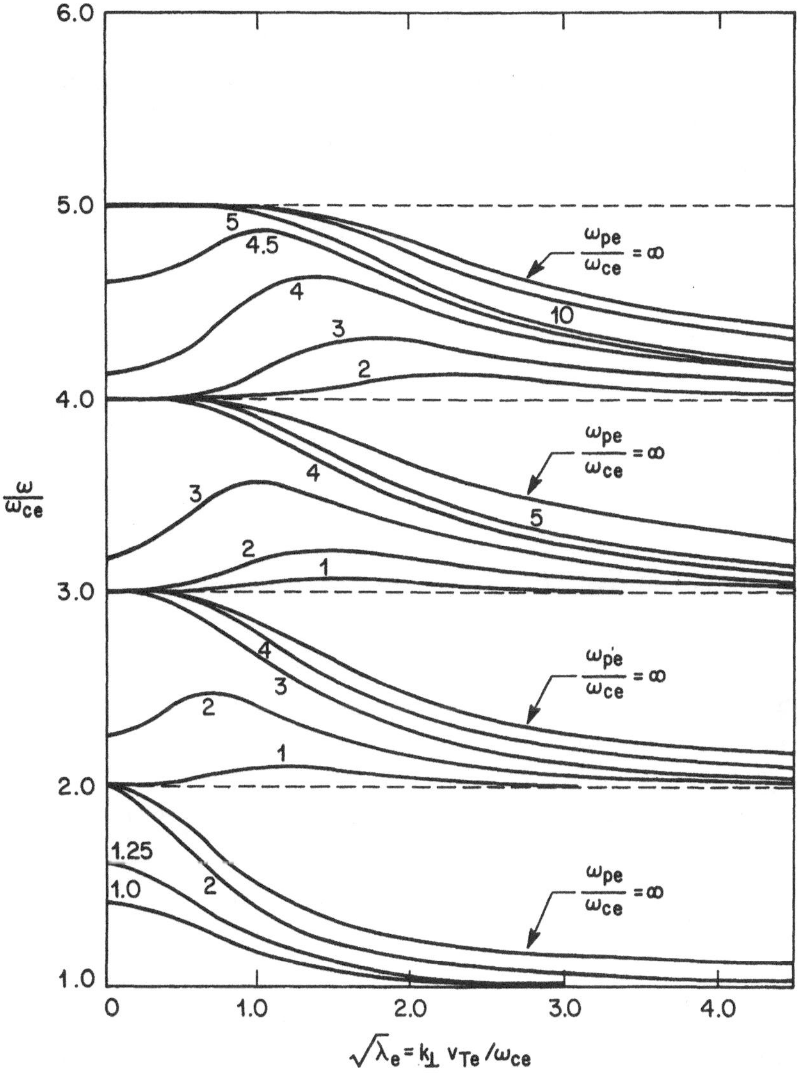

Fig. 2.5. Dispersion relation of the Bernstein wave (G. Bekefi, 1966)

harmonics of the cyclotron frequency are a unique feature of a Vlasov plasma. This feature is called the finite Larmor radius effect because it appears due to the fact that the perpendicular wave length becomes comparable to the Larmor radius of the thermal particles.

When the ion density perturbation is included in Poisson's equation, a similar structure appears at the harmonics of the ion cyclotron frequencies. These waves are called the ion Bernstein waves.

If we include the magnetic field perturbation in the Vlasov equation, we can derive the full electromagnetic plasma dielectric tensor of a Vlasov plasma

$$\overleftrightarrow{\epsilon} = \mathbf{I} - \sum_{\text{species}} \frac{\omega_p^2}{\omega^2}\{\mathbf{I} + \sum_{n=-\infty}^{\infty} \int d\mathbf{v} \frac{k_\parallel(\partial f_o/\partial v_\parallel) + (n\omega_c/v_\perp)(\partial f_o/\partial v_\perp)}{k_\parallel v_\parallel - (\omega - n\omega_c)} \mathbf{S}\}, \tag{2.6.35}$$

where $\mathbf{I}$ is the unit tensor and the tensor $\mathbf{S}$ is given by S =

$$\begin{array}{ccc} \hat{x} & \hat{y} & \hat{z} \end{array}$$

$$\begin{bmatrix} (\frac{n\omega_c}{k_\perp}J_n)^2 & i\frac{n\omega_c}{k_\perp}v_\perp J_n J_n' & \frac{n\omega_c}{k_\perp}v_\parallel J_n^2 \\ -i\frac{n\omega_c}{k_\perp}v_\perp J_n J_n' & (v_\perp J_n')^2 & -iv_\perp v_\parallel J_n J_n' \\ \frac{n\omega_c}{k_\perp}v_\parallel J_n^2 & iv_\perp v_\parallel J_n J_n' & (v_\parallel J_n)^2 \end{bmatrix} \begin{pmatrix} \hat{x} \\ \hat{y} \\ \hat{z} \end{pmatrix}. \tag{2.6.36}$$

The argument of the Bessel functions J_n is $k_\perp v_\perp/\omega_c$; $\omega_c(=qB_o/m)$ is the cyclotron frequency with sign included. In Eq. (2.6.36), $\mathbf{B}_o$ is taken to be in the direction of positive z while the $\mathbf{k}$ vector is taken in the x-z plane. The dispersion relation is obtained by substituting Eq. (2.6.35) into Maxwell's equation,

$$\mathbf{k}(\mathbf{k}\cdot\mathbf{E}_k) - k^2\mathbf{E}_k + \frac{\omega^2}{c^2}\overleftrightarrow{\epsilon}\cdot\mathbf{E}_k = 0. \tag{2.6.37}$$

2.7 Surface Waves

When we take into account a plasma inhomogeneity such as the plasma boundary, a new set of waves emerges. There are two types of waves which propagate uniquely in an inhomogeneous plasma. One is a global wave which is carried by a source at the discontinuous boundary. Such a wave is called a surface wave. The other is a local microscopic wave which is localized in the inhomogeneous region only. This type of wave is called a local mode. Drift waves are typical examples of the local mode.

In this section we describe the surface wave and its nature.

Let us first consider an electrostatic surface wave. We take a plasma density profile which jumps from zero to n_0 at $x = 0$ as shown by the solid line in Fig. 2.6. The potential perturbation which propagates along this surface may be written as

$$\phi(x, y, t) = \phi_k(x)e^{i(ky-\omega t)} + c.c., \tag{2.7.1}$$

where y is a coordinate in a direction along the surface. For simplicity we assume the plasma to be cold, whereby the plasma dielectric constant becomes constant in space. Using $E = -\nabla\phi$ and taking the divergence of Maxwell's equation (2.3.3), we have

$$\nabla\cdot(\overleftrightarrow{\epsilon}\cdot\nabla\phi) = 0, \tag{2.7.2}$$

where $\nabla = \dfrac{\partial}{\partial x}\hat{x} + ik_y$. Since $\overleftrightarrow{\epsilon}$ is constant everywhere except at $x = 0$, Eq. (2.7.2) gives

$$\nabla^2\phi_1 = 0 \quad \text{at } x \neq 0. \tag{2.7.3}$$

For the assumed structure of ϕ of Eq. (2.7.1), the solution of Eq. (2.7.3) is given by

$$\phi_k(x) = \phi_b e^{-\kappa|x|}, \tag{2.7.4}$$

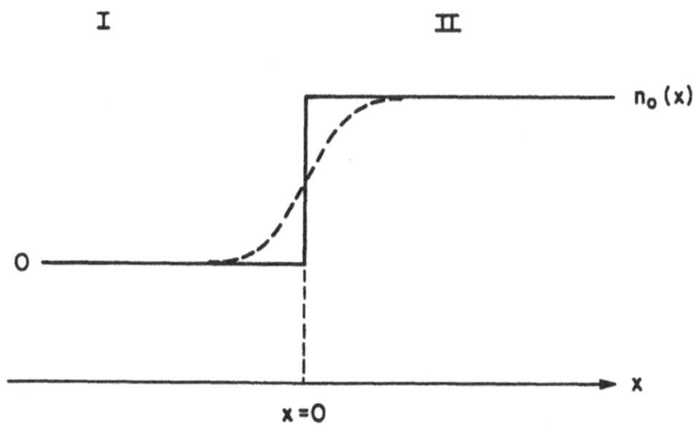

Fig. 2.6. Plasma density profile for a surface wave

where $\kappa = |k|$ and ϕ_b is the potential at the boundary. The field given by Eq. (2.7.4) has a discontinuous peak at $x=0$ and decays away from the boundary; hence the associated wave is called a surface wave. The dispersion relation of the surface wave is obtained by using the boundary conditions that D_n (the normal component of the displacement) and ϕ are continuous across the boundary. Since the continuity of ϕ is satisfied for the choice of the solution (2.7.4), the continuity of $D_n = \epsilon_{xx}\partial\phi/\partial x$ gives the dispersion relation of the surface wave

$$\epsilon_{\perp I} + \epsilon_{\perp II} = 0, \tag{2.7.5}$$

where subscripts I and II refer to quantities on each side of the boundary and $\epsilon_\perp$ is the xx component of the dielectric constant. In particular if side I is a vacuum, $\epsilon_\perp = 1$, while if side II is a plasma with a magnetic field in the z direction, $\epsilon_\perp = 1 - \omega_{pe}^2/(\omega^2 - \omega_{ce}^e)$. Then the dispersion relation becomes,

$$\omega^2 = \frac{\omega_p^2}{2} + \omega_c^2 . \tag{2.7.6}$$

Furthermore if there exists no magnetic field, the surface wave frequency is given by $\omega_p/\sqrt{2}$ which is smaller than the plasma frequency. The surface wave is carried along the surface by the induced charge at the boundary surface.

An important surface wave in space plasmas is the *surface Alfveń wave.* The surface Alfveń wave may be studied using the linearized MHD equations introduced in Section 2.5. We again take a plane and sharp boundary at $x=0$ and designate quantities in the $x<0$ region by subscript I and $x>0$ region by subscript II. We assume that the magnetic field is straight and directed in the z direction. The background plasma density, magnetic field and pressure are discontinuous at the boundary.

On both sides of the boundary, where all the background quantities are constant, the equation of motion (2.5.1a) can be reduced by the help of (2.5.4a), (2.5.5a) and (2.5.6a) to

$$m_i n_o \frac{\partial^2 \mathbf{v}_1}{\partial t^2} - \frac{1}{\mu_o}(\mathbf{B}_o \cdot \nabla)^2 \mathbf{v}_1 = -\nabla \frac{\partial \tilde{p}_1}{\partial t}, \tag{2.7.7}$$

where $\tilde{p}_1$ is the perturbation of the total plasma pressure given by

$$\tilde{p}_1 = p_1 + \frac{\mathbf{B}_o \cdot \mathbf{B}_1}{\mu_o}. \tag{2.7.8}$$

The field equation which represents the surface Alfveń wave is obtained by taking the divergence of Eq. (2.7.7) and by assuming an incompressible mode $\nabla \cdot \mathbf{v}_1 = 0$, as

$$\nabla^2 \tilde{p}_1 = 0. \tag{2.7.9}$$

As in the case of the electrostatic surface wave, the surface wave equation for the Alfveń wave is also given by the Laplace equation. This is a common feature of surface waves and represents the fact that the source is localized at the boundary. For a perturbation of the form

$$\tilde{p}_1(x, y, z, t) = p_k(x) e^{i(k_\parallel z + k_\perp y - \omega t)} + \text{c.c.} \tag{2.7.10}$$

the solution of Eq. (2.7.8) reads

$$\tilde{p}_k = \tilde{p}_b e^{-\kappa |x|}, \tag{2.7.11}$$

where $\kappa = (k_\parallel^2 + k_\perp^2)^{1/2}$. The dispersion relation is obtained by using the boundary condition that $\tilde{p}_1$ and the normal component of $\mathbf{v}_1 \equiv v_{1x}$ are continuous,

$$\omega^2 = \frac{(B_{oI}^2 + B_{oII}^2)}{\mu_o m_i (n_{oI} + n_{oII})} k_\parallel^2 . \tag{2.7.12}$$

In particular if side I is a vacuum, $n_{oI} = 0$ and if $B_{oI} = B_{oII} \equiv B_o$, the surface Alfveń wave dispersion relation is given by

$$\omega^2 = \frac{2B_o}{\mu_o m_i n_o} k_\parallel^2 = 2k_\parallel^2 v_A^2 . \tag{2.7.13}$$

We note that, in contrast to the electrostatic surface wave where the wave frequency $(\omega_p/\sqrt{2})$ was smaller than the bulk mode frequency (ω_p), the surface Alfveń wave frequency is larger than the bulk Alfveń frequency. However, it is important to recognize the fact that the surface wave frequency always exists between the bulk mode frequencies of both sides of the boundary. This fact indicates that when the boundary is smooth, as shown by the dotted line in Fig. 2.6, the surface wave frequency matches the local bulk mode frequency somewhere on the smooth boundary. For example, if region I is a vacuum, the bulk Alfveń wave frequency will approach infinity (speed of light) in region I and will decrease smoothly to $k_\parallel v_A$ at region II. Hence, there exists a local surface where the local Alfveń frequency $k_\parallel v_A(x)$ will become the same as the surface Alfveń wave frequency of $\sqrt{2} k_\parallel v_A$. In the next Section we discuss the resonant absorption of the surface wave caused by this local resonance.

2.8 Resonant Absorption and the Kinetic Alfvén Wave

Resonant Absorption

Let us consider the bulk Alfvén wave which propagates in the smooth boundary regions where the background quantities vary as a function of x. Since B_o is now a function of x, the equilibrium condition requires that an equilibrium current density $\mathbf{J}_o(x)$ exist in the y direction. Using this fact, the linearized equation of motion with the help of Ampere's Law becomes

$$m_i n_o \frac{\partial^2 \mathbf{v}_1}{\partial t^2} = \frac{1}{\mu_o}[(\mathbf{B}_o \cdot \nabla)\frac{\partial \mathbf{B}_1}{\partial t} + (\frac{\partial \mathbf{B}_1}{\partial t} \cdot \nabla)\mathbf{B}_o] - \nabla \frac{\partial \tilde{p}_1}{\partial t} \quad (2.8.1)$$

for an incompressible wave ($\nabla \cdot \mathbf{v}_1 = 0$). Here, from Eqs. (2.5.4a) and (2.5.2a),

$$\frac{\partial \mathbf{B}_1}{\partial t} = (\mathbf{B}_o \cdot \nabla)\mathbf{v}_1 - (\mathbf{v}_1 \cdot \nabla)\mathbf{B}_o . \quad (2.8.2)$$

If we now use the fact that $\mathbf{B}_o = B_o(x)\hat{\mathbf{z}}$ and that $\mathbf{v}_1$ may be written as

$$\mathbf{v}_1(x, y, z, t) = \mathbf{v}_k(x) e^{i(k_\perp y + k_\parallel z - \omega t)} + \text{c.c.} \quad (2.8.3)$$

Eqs. (2.8.1) and (2.8.2) may be reduced to

$$(m_i n_o \omega^2 - \frac{B_o^2 k_\parallel^2}{\mu_o})\mathbf{v}_k = \nabla \frac{\partial \tilde{p}_k}{\partial t}, \quad (2.8.4)$$

where $\nabla = ik_\perp \hat{\mathbf{y}} + ik_\parallel \hat{\mathbf{z}} + (\partial/\partial x)\hat{\mathbf{x}}$. To eliminate $\tilde{p}_k$, we take the curl of both sides to give

$$\nabla \epsilon_A \times \mathbf{v}_k + \epsilon_A \nabla \times \mathbf{v}_k = 0, \quad (2.8.5)$$

$$\text{where} \quad \epsilon_A(x) = m_i n_o(x)\omega^2 - \frac{B_o^2(x) k_\parallel^2}{\mu_o}. \quad (2.8.6)$$

From the x-component of Eq. (2.8.5) we see that $(\nabla \times \mathbf{v}_k) \cdot \hat{\mathbf{x}} = 0$, hence the y and z components of $\mathbf{v}_k$ are related through

$$k_\perp v_{zk} - k_\parallel v_{yk} = 0. \quad (2.8.7)$$

The z-component of Eq. (2.8.5) gives

$$\frac{d\epsilon_A}{dx}v_{yk} + \epsilon_A\left(\frac{\partial v_{yk}}{\partial x} - ik_\perp v_{xk}\right) = 0, \qquad (2.8.8)$$

while the incompressibility condition gives

$$\frac{\partial v_{xk}}{\partial x} + ik_\perp v_{yk} + ik_\parallel v_{zk} = 0. \qquad (2.8.9)$$

If we eliminate v_y and v_z from Eqs. (2.8.7) to (2.8.9), Eq. (2.8.8) finally reads

$$\frac{d}{dx}\left(\epsilon_A \frac{dv_{xk}}{dx}\right) - \epsilon_A(k_\perp^2 + k_\parallel^2)v_{xk} = 0. \qquad (2.8.10)$$

This is the desired wave equation for an Alfveń wave propagating in an inhomogeneous region of the plasma. We note that if $\epsilon_A = \text{const}$, Eq. (2.8.10) admits two types of solution, one, $\epsilon_A = 0$ which gives the bulk Alfveń wave dispersion relation, and the other, $\nabla^2 v_x = 0$ which gives the surface wave. Hence Eq. (2.8.10) may be considered as representing the coupling between the local Alfveń wave and the Alfveń surface wave.

As an example, let us consider a case where region I is a vacuum. Then the surface Alfveń wave has the eigen frequency given by

$$\omega = \omega_s = \sqrt{2}k_\parallel v_A, \qquad (2.8.11)$$

where v_A represents the Alfveń speed in the region II. Now for this eigen frequency $\omega = \omega_s$, there exists a position $x = x_0$ where $\epsilon_A(x_0)$ vanishes in the boundary region, because $\epsilon_A(x \to -\infty) = -B_o^2k_\parallel^2/\mu_o$ while $\epsilon_A(x \to +\infty) = 2m_in_0k_\parallel^2v_A^2 - B_o^2k_\parallel^2/\mu_o = k_\parallel^2B_o^2/\mu_o$. At this point, the wave equation (2.8.10) becomes singular since the coefficient of d^2v_{xk}/dx^2 vanishes. The solution near the singular point $x = x_0$ may be obtained by expanding ϵ_A near x_0,

$$\epsilon_A(x) = \epsilon_A(x = x_0) + (x - x_0)\epsilon_A' + i0,$$

where $\epsilon_A' = d\epsilon_A/dx$ at $x = x_0$ and use is made of the fact $\text{Im}\,\omega = +0$. Then the solution of Eq. (2.8.10) can be written

$$\left.\begin{aligned} v_{xk} &= \frac{c}{\epsilon_A'}\ell n(x - x_0) && x > x_o \\ &= \frac{c}{\epsilon_A'}(\ell n|x - x_0| + i\pi\,\text{sign}\,\epsilon_A'), && x < x_0 \end{aligned}\right\}, \qquad (2.8.12)$$

where c is the integration constant. The solution across the singular surface x_o is connected through analytic continuation using the fact that $\mathrm{Im}\omega = +0$. The singular behavior of the solution indicates absorption of the surface wave energy. The absorption can be identified by evaluating the power flow across the singular surface. The power flow density P in the x direction is given by

$$P = \frac{1}{2} R_e(\tilde{p}_k v_{xk}^*). \tag{2.8.13a}$$

From the y-component of Eq. (2.8.4), we have

$$\epsilon_A v_{yk} = k_\perp \omega_s \tilde{p}_k \tag{2.8.14}$$

while Eq. (2.8.9) gives, for $k_\perp >> k_\parallel$,

$$i k_\perp v_{yk} + \frac{\partial v_{xk}}{\partial x} = 0. \tag{2.8.15}$$

Eliminating v_{yk} from Eqs. (2.8.14) and (2.8.15) we have

$$\tilde{p}_k = \frac{i\epsilon_A}{k_\perp^2 \omega_s} \frac{\partial v_{xk}}{\partial x} \tag{2.8.16}$$

Thus the power flow Eq. (2.8.13) becomes

$$\left.\begin{aligned} P &= \frac{c^2}{2k_\perp^2} \frac{\pi}{\omega_s |\epsilon_A'|} && \text{for } x < x_o \\ &= 0 && \text{for } x > x_o \end{aligned}\right\}. \tag{2.8.13b}$$

The fact that there exists a jump of x-directed power flow between $x < x_o$ and $x > x_o$ indicates a local absorption of the surface wave energy at the local Alfveń resonance layer $x = x_o$, where ϵ_A vanishes, i.e.,

$$m_i n_o(x_o)\omega_s^2 - \frac{B_o^2(x_o)k_\parallel^2}{\mu_o} = 0. \tag{2.8.17}$$

The absorption of the wave energy in an ideal MHD situation may seem strange. The absorption is a consequence of the mathematical singularity existing in the MHD equation which is associated with the fact that the Alfveń wave propagates only in the direction of the magnetic field. The singularity originates from the assumption of zero-Larmor radius used in the derivation of the MHD equations.

Kinetic Alfvén Wave

In order to avoid the singularity, one must include the finite Larmor radius effect in MHD equations. A simple way to do so is to use the generalized Ohm's law Eq. (1.6.13). Then the first term produces a correction of the order of $k_\perp c/\omega_{pe}$, while the second term of the order of $k_\perp \rho_s = k_\perp \sqrt{T_e/m_i}\,/\omega_{ci}$. Furthermore, the existence of these terms indicates the appearance of a parallel electric field. However these corrections do not properly account for the appearance of dissipation due to the Landau damping. To illustrate this, we use here the Vlasov equations. The parallel electric field which is generated by the finite Larmor radius effect is expected to let the Alfvén wave couple with the ion acoustic wave. However, if $k_\perp \rho_i \simeq O(1)$, the frequency of the fast (compressional) mode becomes of the order of the ion-cyclotron frequency; hence we can consider that the fast mode is decoupled. Then the compressional component of the magnetic-field perturbation, B_z, can be assumed to be much smaller than the transverse components. This allows us to use a scalar potential, ϕ, to represent the transverse components of the electric field, $\mathbf{E}_\perp$,

$$\mathbf{E}_\perp = -\nabla_\perp \phi \tag{2.8.18}$$

because then B_z becomes zero. To represent the z component of the electric field, we must use a different potential, ψ,

$$E_z = -\frac{\partial \psi}{\partial z} \tag{2.8.19}$$

so that the transverse components of $\nabla \times \mathbf{E}$ are not zero. The appropriate field equations are Poisson's equation,

$$\nabla_\perp^2 \phi + \frac{\partial^2 \psi}{\partial z^2} = \frac{e}{\epsilon_0}[n^{(i)} - n^{(e)}] \approx 0 \tag{2.8.20}$$

and the z component of Ampere's law,

$$\frac{\partial}{\partial z}\nabla_\perp^2(\phi - \psi) = \mu_0 \frac{\partial}{\partial t}[J^{(i)} + J^{(e)}]. \tag{2.8.21}$$

The quantities $n^{(i)}$, $n^{(e)}$, $J^{(i)}$ and $J^{(e)}$ are obtained from the linearized Vlasov equation for each species. For electrons, the linearized distribution function $f_k^{(e)}$ becomes, assuming $k_\perp \rho_e \simeq 0$ and $\omega << \omega_{ce}$,

$$f_k^{(e)}(\mathbf{v}) = -\frac{e}{m_e}\frac{k_{\parallel}\psi_k}{k_{\parallel}v_{\parallel} - \omega}\frac{\partial f_0^{(e)}}{\partial v_{\parallel}}. \tag{2.8.22}$$

For ions, by retaining the finite Larmor radius effect,

$$f_k^{(i)}(v)=f_k^{(i)}(v_\perp,\theta,v_\parallel)=-\frac{e}{m_i}\sum_{n=-\infty}^{\infty}\sum_{m=-\infty}^{\infty}\frac{e^{i(n-m)\theta}}{\omega-k_\parallel v_\parallel-n\omega_{ci}}J_n\left(\frac{k_\perp v_\perp}{\omega_{ci}}\right)J_m\left(\frac{k_\perp v_\perp}{\omega_{ci}}\right)$$

$$\left[k_\perp\phi_k\left\{\left(1-\frac{k_\parallel v_\parallel}{\omega}\right)\frac{n\omega_{ci}}{k_\perp v_\perp}\frac{\partial f_o^{(i)}}{\partial v_\perp}+\frac{k_\parallel}{k_\perp}\frac{n\omega_{ci}}{\omega}\frac{\partial f_o^{(i)}}{\partial v_\parallel}\right\}\right.$$

$$\left.+k_\parallel\psi_k\left\{\frac{v_\parallel}{v_\perp}\frac{n\omega_{ci}}{\omega}\frac{\partial f_o^{(i)}}{\partial v_\perp}+\left(1-\frac{n\omega_{ci}}{\omega}\right)\frac{\partial f_o^{(i)}}{\partial v_\parallel}\right\}\right], \qquad (2.8.23)$$

where θ is the phase angle of the Larmor motion. The Fourier amplitudes of the charge density and the current density perturbation are obtained from

$$qn_k^{(j)}=qn_o\int f_k^{(j)}dv \qquad j=1,\ e \qquad (2.8.24)$$

$$J_k^{(j)}=qn_o\int v f_k^{(j)}dv \qquad j=1,\ e. \qquad (2.8.25)$$

If the plasma beta is larger than m_e/m_i, the electron thermal speed will become larger than the Alfvén speed and the resultant charge and current density perturbations may be reduced to

$$\frac{en_k^{(i)}}{\epsilon_o}=-\frac{\omega_{pi}^2}{v_{Ti}^2}[1-I_o(\lambda_i)e^{-\lambda_i}]\phi_k+\frac{\omega_{pi}^2k_\parallel^2}{\omega^2}I_o(\lambda_i)e^{-\lambda_i}(1-i\delta_i)\psi_k \qquad (2.8.26)$$

$$\frac{en_k^{(e)}}{\epsilon_o}=\frac{\omega_{pe}^2}{v_{Te}^2}(1+i\delta_e)\psi_k \qquad (2.8.27)$$

$$\mu_o J_{zk}^{(i)}=\frac{\omega_{pi}^2}{c^2}\frac{k_\parallel}{\omega}I_o(\lambda_i)e^{-\lambda_i}(1-i\delta_i)\psi_k \qquad (2.8.28)$$

$$\mu_o J_{zk}^{(e)}=-\frac{\omega_{pe}^2}{c^2v_{Te}^2}\frac{\omega}{k_\parallel}(1+i\delta_e)\psi_k, \qquad (2.8.29)$$

where $\lambda_i=k_\perp^2\rho_i^2$, $I_o(\lambda_i)$ is the modified Bessel function of the first kind, and δ_i and δ_e are the fractional Landau damping rates,

$$\delta_i=2(\pi)^{1/2}\beta_i^{-3/2}\exp(-\beta_i^{-1}) \qquad (2.8.30)$$

$$\delta_e=(\pi)^{1/2}\beta_i^{-1/2}\left(\frac{T_i}{T_e}\right)^{1/2}\left(\frac{m_e}{m_i}\right)^{1/2}. \qquad (2.8.31)$$

Here,

$$\beta_i = 2\frac{v_{Ti}^2}{v_A^2} \tag{2.8.32}$$

and v_{Te} and v_{Ti} are the electron and ion thermal speeds, respectively. The dispersion relation can be obtained by substituting Eqs. (2.8.26) to (2.8.29) into (2.8.20) and (2.8.21). If we ignore the damping,

$$(I_0 e^{-\lambda_i} - \frac{\omega^2}{k_\parallel^2 c_s^2})[1 - \frac{\omega^2}{k_\parallel^2 v_A^2}\frac{1}{\lambda_i}(1 - I_0 e^{-\lambda_i})] = \frac{\omega^2}{k_\parallel^2 v_{Ti}^2}(1 - I_0 e^{-\lambda_i}), \tag{2.8.33}$$

where c_s is the ion sound speed with electron temperature, $c_s^2 = T_e/m_i$. This dispersion relation shows the coupling of the Alfveń wave and the ion acoustic wave. In a low-beta plasma, since $v_s^2 << v_A^2$, the two waves are decoupled, and the dispersion relation of the Alfveń wave becomes

$$\frac{\omega^2}{k_\parallel^2 v_A^2} = \frac{\lambda_i}{1 - I_0 e^{-\lambda_i}} + \frac{T_e}{T_i}\lambda_i . \tag{2.8.34}$$

Here we call the wave represented by this dispersion relation the "kinetic Alfveń wave" because of its kinetic property. If $\lambda_i << 1$, the dispersion relation of the kinetic Alfveń wave reduces to

$$\omega^2 = k_z^2 v_A^2 \left[1 + k_\perp^2 \rho_i^2 \left(\frac{3}{4} + \frac{T_e}{T_i}\right)\right]. \tag{2.8.35}$$

We note that, unlike the MHD Alfveń wave, the kinetic Alfveń wave can propagate across the magnetic field and will undergo both electron and ion Landau damping because of its coupling to the electrostatic mode. It is also important to note that the wave has an electric field component in the direction of the ambient magnetic field.

If the plasma is relatively cold such that $v_{Te} < v_A$, the electron inertia becomes important and the dispersion relation is modified to

$$\omega^2 = k_\parallel^2 v_A^2 (1 + \frac{3}{4} k_\perp^2 \rho_i^2) \left(1 + \frac{k_\perp^2 c^2}{\omega_{pe}^2}\right)^{-1} . \tag{2.8.36}$$

We also note that for $T_i = 0$ but $T_e \neq 0$ the dispersion relation (2.8.35) and (2.8.36) may be obtained from the MHD equations if one uses the modified Ohm's law, Eq. (1.6.13).

Resonant Mode Conversion

Expecting that the finite Larmor radius effect will eliminate the singularity, we now derive the wave equation for the Alfveń wave in the inhomogeneous region between I and II in Fig. 2.6 using the Vlasov equation. To take into account the plasma density inhomogeneity, we must use the equilibrium distribution function given by

$$f_o(x, \mathbf{v}) = g\left(x + \frac{v_y}{\omega_c}\right) f_o(\mathbf{v}) \tag{2.8.37}$$

where $f_o(\mathbf{v})$ is the Maxwellian distribution. The perturbed distribution function is expressed as

$$f_1(\mathbf{x}, \mathbf{v}, t) = f_k(x) e^{i(k_y y + k_{\parallel} z - \omega t)} + \text{c.c.} \tag{2.8.38}$$

To study the resonant mode conversion of the Alfveń wave, we can assume that k_y, the wave number perpendicular to the magnetic field as well as to the density gradient, is much smaller than ρ_i^{-1} and that the scale size of the density gradient is much larger than the ion gyroradius. Thus,

$$\frac{\omega}{\omega_{ci}} >> k_y \kappa \rho_i^2 ,$$

where κ^{-1} is the inhomogeneity scale. This assumption allows us to ignore the term v_y/ω_{ci} in g in Eq. (2.8.37), which is equivalent to neglecting a drift wave as will be discussed in Section 2.9.

In addition, we assume that the wavelength in the x direction near the mode conversion point is small but larger than the ion gyroradius; we can then expand the wave equation in the power of $\rho_i d/dx$. We also assume $v_{Te} > v_A$.

With these assumptions the Fourier amplitude of the number density and the current-density perturbations are given by

$$\frac{e n_k^{(i)}}{\epsilon_o} = \frac{\omega_{pi}^2}{\omega_{ci}^2}\left[\frac{d}{dx}\left(1 + \frac{3}{4}\rho_i^2 \frac{d^2}{dx^2}\right)\left(g\frac{d\phi_k}{dx}\right) - g k_y^2 \phi_k\right] + \frac{\omega_{pi}^2}{\omega^2} k_{\parallel}^2 g \psi_k \tag{2.8.39}$$

$$\frac{e n_k^{(e)}}{\epsilon_o} = \frac{\omega_{pe}^2}{v_{Te}^2} g \psi_k \tag{2.8.40}$$

$$\mu_o J_{zk}^{(i)} = \frac{\omega_{pi}^2}{c^2}\frac{k_{\parallel}}{\omega} g\psi_k \tag{2.8.41}$$

and

$$\mu_o J_{zk}^{(e)} = -\frac{\omega_{pe}^2}{c^2 v_{Te}^2}\frac{\omega}{k_{\parallel}} g\psi_k . \tag{2.8.42}$$

In these expressions the ion and the electron plasma frequencies, ω_{pi} and ω_{pe}, respectively, are constants and refer to the values at the maximum plasma density, where g(x) is normalized to unity.

To study the resonant mode conversion to the kinetic Alfve wave, we consider only the region near the resonant point $x = x_o$, where $\omega^2 = k_{\parallel}^2[v_A^2/g(x=x_o)]$. We can then further simplify the expressions by using the assumption of a low-beta plasma so that $v_A^2 >> c_s^2 = T_e/m_i$. This assumption eliminates the possibility of a simultaneous coupling to the ion acoustic wave. We can then derive the wave equation by eliminating ψ from Eqs. (2.8.20), (2.8.21) and (2.8.39) to (2.8.42),

$$\left(\frac{\omega^2}{k_{\parallel}^2 v_A^2}\frac{3}{4}\rho_i^2\frac{d^3}{dx^3} + \frac{d^2}{dx^2}\frac{1}{g}\frac{T_e}{T_i}\rho_i^2\frac{d}{dx}\right)\left(g\frac{d\phi_k}{dx}\right)$$
$$+\left[\frac{d}{dx}\left(\frac{\omega^2}{k_{\parallel}^2 v_A^2}g-1\right)\frac{d}{dx} - k_y^2\left(\frac{\omega^2}{k_{\parallel}^2 v_A^2}-1\right)\right]\phi_k = 0, \tag{2.8.43}$$

where the Alfveń speed, v_A^2, is that of the maximum density and g is normalized to unity.

We can immediately notice that, if we put $\rho_i \rightarrow 0$ in this expression, the wave equation reduces to

$$\frac{d}{dx}\left[\epsilon(x)\frac{d\phi_k}{dx}\right] - k_y^2\epsilon(x)\phi_k = 0, \tag{2.8.44}$$

where $\epsilon(x) = (\omega^2/k_{\parallel}^2 v_A^2)g - 1$. Equation (2.8.44) has a structure identical to the MHD wave Eq. (2.8.10). We also note that in a uniform plasma, $g = 1$, Eq. (2.8.43) gives two decoupled wave equations

$$\nabla_{\perp}^2\phi_k = 0 \tag{2.8.45}$$

and

$$\left[\bar{\rho}^2\frac{d^2}{dx^2}+\left(\frac{\omega^2}{k_{\parallel}^2 v_A^2}-1\right)\right]\phi_k=0, \tag{2.8.46}$$

where $\bar{\rho}^2=[(3/4)+(T_e/T_i)]\rho_i^2$. The assumption $k_y << d/dx$ is used in (4.8.46).

Equation (4.8.45) represents a quasi-static electromagnetic perturbation (a cutoff mode) associated with an external source. In the absence of a source, this equation represents a surface wave. Equation (2.8.46) is the wave equation for the bulk kinetic Alfveń wave, Eq. (2.8.35).

We can thus deduce that Eq. (2.8.43) represents a coupling between a surface MHD mode or an externally applied electromagnetic perturbation and the kinetic Alfveń wave. From Eq. (2.8.46) we can see that the kinetic Alfveń wave propagates, after the mode conversion, to the higher density side, where $k_{\parallel}^2 v_A^2(x) < k_{\parallel}^2 v_A^2(x=x_o)$.

To study the mode conversion, we must specify the actual density profile. As a simple example, we take a linear profile for the plasma density so that

$$g(x)=\kappa x+a,$$

where x is a normalized distance with the origin being located at the resonant point where $g(x=0)\omega^2/(k_{\parallel}^2 v_A^2)=1$, or $a\omega^2/k_{\parallel}^2 v_A^2=1$, with $0<a<1$. Equation (2.8.43) is then reduced near $x\approx 0$ to the simple form

$$\rho^2\frac{d^2E_x}{dx^2}+\kappa x E_x=E_o, \tag{2.8.47}$$

where

$$\rho^2=\left(\frac{3}{4}+\frac{k_{\parallel}^2 v_A^2}{\omega^2}\frac{T_e}{T_i}\right)\rho_i^2 \tag{2.8.48}$$

and

$$E_x=-\frac{\partial\phi}{\partial x}. \tag{2.8.49}$$

E_o is an integration constant representing a nominal value of E_x at a large negative x (A value of E_x associated with the external source field or surface wave).

In the derivation we also assumed that $|(dg/dx)/g| << |(d\phi_k/dx)/\phi_k|$; i.e., the variation of the wave amplitude is much faster than the variation of the density (W.K.B. approximation). The solution can be expressed in terms of the Airy functions.

If we introduce a scale length.

$$\Delta = (\frac{\rho^2}{\kappa})^{1/3} \tag{2.8.50}$$

and use a normalized electric-field intensity,

$$\bar{E}_x = -\frac{E_x}{E_o}\frac{(\kappa\rho)^{2/3}}{\pi}, \tag{2.8.51}$$

the general solution is given by

$$\bar{E}_x = c_1 A_i(-x/\Delta) + c_2 B_i(-x/\Delta) + G_i(-x/\Delta), \tag{2.8.52}$$

where A_i and B_i are Airy functions and G_i is a function involving integrals of A_i and B_i.

The integration constants c_1 and c_2 can be determined by applying appropriate boundary conditions. For a semi-infinite plasma as considered here, they can be determined whether or not the energy source exists internally or externally to the plasma.

For resonant absorption or surface-wave damping, the suitable boundary condition for the kinetic Alfveń wave is to accept only the right-going wave (no reflection at $x \rightarrow \infty$) and that which has no divergence at $x \rightarrow -\infty$. We can then find $c_2 = 0$ and $c_1 = i$. The asymptotic solution for $|x/\Delta| >> 1$ can then be written as

$$E_x = -\frac{\pi^{1/2} E_o}{(\kappa\rho)^{2/3}}\left(\frac{\Delta}{x}\right)^{1/4} \exp\left\{i\left[\frac{2}{3}(\frac{x}{\Delta})^{3/2} + \frac{\pi}{4}\right]\right\} + \frac{E_o}{\kappa x} \tag{2.8.53}$$

for $x > 0$, and

$$E_x = \frac{E_o}{\kappa x} \tag{2.8.54}$$

for $x < 0$. The first term in Eq. (2.8.53) represents the kinetic Alfveń wave and the second term, as well as the expression in Eq. (2.8.54), the source or surface-wave field.

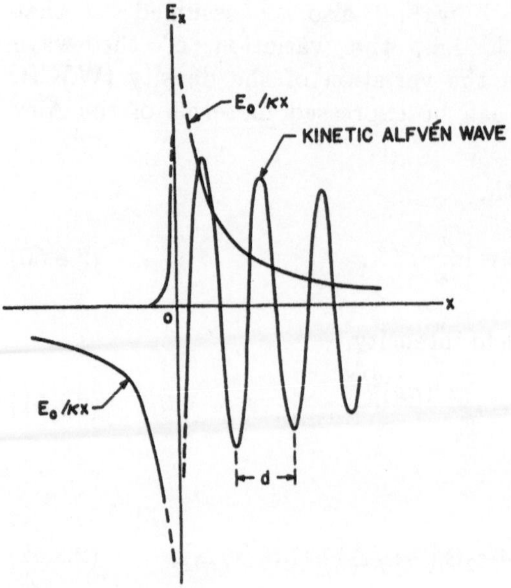

Fig. 2.7. Mode-conversion to the kinetic Alfveń wave

As is expected, the kinetic Alfveń wave propagates on the higher density side ($x \geq 0$) of the resonant point, as shown in Eq. (2.8.53). In Eq. (2.8.53) the first few peak amplitudes of the kinetic Alfveń wave after the mode conversion are given by $E_o(\kappa\rho)^{-2/3}$, with an effective wave number of $(\kappa/\rho^2)^{1/3}$, whereas away from the resonant point, say at $x \approx \kappa^{-1}$, the amplitude and the wave number of the kinetic Alfveń wave become $E_o(\kappa\rho)^{-1/2}$ and ρ^{-1}, respectively. The qualitative feature of the mode-converted kinetic Alfveń wave is shown in Fig. 2.7.

We can see that the solution for $x < 0$ is identical to that obtained under the ideal MHD approximation. Hence the plasma impedance, and consequently the absorption rate, observed from the surface wave remains unchanged from the previous MHD calculation.

If the energy source exists inside the plasma, we must take into account the left-going wave as well. If $v_{Te} < v_A$ at the resonance surface, x_o, the appropriate dispersion relation of the kinetic Alfveń wave is approximately given by

$$\omega^2 = k_\parallel^2 v_A^2 \left(1 + \frac{c^2 k_\perp^2}{\omega_{pe}^2}\right)^{-1} . \tag{2.8.55}$$

Hence, after the mode conversion, the kinetic wave propagates toward the lower density side in this case.

2.9 Drift Wave

The drift wave is a local wave which propagates in the direction of the diamagnetic drift velocity (explained later) in a inhomogeneous portion of the plasma. The nature of the drift wave may be most simply illustrated by considering an electrostatic perturbation in a plasma with $T_i \ll T_e$. Here one can treat the ions as a cold fluid while the electrons obey an isothermal Boltzmann equilibrium.

In an electrostatic field the cold ion fluid may be described by the equation of motion,

$$\frac{d\mathbf{v}}{dt} = -\frac{e}{m_i}\nabla\phi + \omega_{ci}\mathbf{v}\times\hat{\mathbf{z}}, \tag{2.9.1}$$

where ϕ is the electrostatic potential of the field, and the static magnetic field is taken in the direction of the z axis. Since the drift wave is a local wave with the perpendicular wavelength $k_\perp^{-1}$ much shorter than the parallel wavelength $k_\parallel^{-1}$, the gradient operative is given approximately by the perpendicular gradient, here $\nabla \simeq \nabla_\perp$. To study a wave which has a short perpendicular wavelength, it is convenient to construct the equation of vorticity. If we take the z-component of the curl of Eq. (2.9.1), we have

$$\frac{d}{dt}(\omega_{ci}+\Omega) + (\omega_{ci}+\Omega)\nabla\cdot\mathbf{v} = 0, \tag{2.9.2}$$

where $\Omega = \hat{\mathbf{z}}\cdot\nabla\times\mathbf{v}$ and use is made of that relation

$$\begin{aligned}\nabla\times(\mathbf{v}\cdot\nabla)\mathbf{v} &= -\nabla\times(\mathbf{v}\times\boldsymbol{\Omega})\\ &= \boldsymbol{\Omega}\nabla\cdot\mathbf{v} + (\mathbf{v}\cdot\nabla)\boldsymbol{\Omega} - (\boldsymbol{\Omega}\cdot\nabla)\mathbf{v}\\ &\simeq \boldsymbol{\Omega}\nabla\cdot\mathbf{v} + (\mathbf{v}\cdot\nabla)\boldsymbol{\Omega}.\end{aligned} \tag{2.9.3}$$

If we combine Eq. (2.9.2) and the equation of continuity,

$$\nabla\cdot\mathbf{v} = -\frac{d}{dt}\ell n\, n, \tag{2.9.4}$$

we can construct the equation of conservation of vorticity for a compressible fluid,

$$\frac{d}{dt}\ell n\left(\frac{\omega_{ci}+\Omega}{n}\right) = 0. \tag{2.9.5}$$

We note that this equation is valid only if the $(\boldsymbol{\Omega}\cdot\nabla)\mathbf{v}$ term is negligible in Eq. (2.9.3) i.e., only for almost two dimensional perturbations.

For a slow vortex motion of the fluid, the vorticity Ω may be expressed in terms of ϕ from Eq. (2.9.1) by dropping the inertial term

$$\mathbf{v}=\frac{-\nabla\phi\times\hat{\mathbf{z}}}{B_o} \tag{2.9.6}$$

and by taking the curl

$$\Omega=\frac{\nabla^2\phi}{B_o}. \tag{2.9.7}$$

The number density of the ions should be the same as for electrons (quasi-neutrality requirement) and is expected to obey the Boltzmann distribution.

$$n=n_0(x)\exp\left(\frac{e\phi}{T_e}\right). \tag{2.9.8}$$

If we substitute Eq. (2.9.6) to (2.9.8) into the equation for vorticity conservation and expand the logarithm in the power of ϕ, we have a closed equation for ϕ,

$$\frac{\partial}{\partial t}(\rho_s^2\nabla_\perp^2\phi-\phi)-\frac{1}{B_o}(\nabla\phi\times\hat{\mathbf{z}})\cdot\nabla\left(\rho_s^2\nabla_\perp^2\phi+\frac{T_e}{e}\ell n\left(\frac{\omega_{ci}}{n_0}\right)\right)=0, \tag{2.9.9}$$

where $\rho_s=c_s/\omega_{ci}=(T_e/m_i)^{1/2}/\omega_{ci}$. To consider a mode localized in the inhomogeneous region, we take a uniform magnetic field and an exponential spatial profile for the density n_0 such that

$$n_0\sim\exp(-\kappa x), \tag{2.9.10}$$

and linearize Eq. (2.9.9) by assuming a local mode profile of the form,

$$\phi(\mathbf{x},t)=\phi_k e^{i(k_y y-\omega t)}+\text{c.c.}, \tag{2.9.11}$$

where c.c. again is the complex conjugate. Equation (2.9.9) then gives the dispersion relation for the local mode,

$$\omega=\frac{\omega_{*e}}{1+k_\perp^2\rho_s^2}, \tag{2.9.12}$$

where ω_{*e} is called the electron drift wave frequency defined as

$$\omega_{*e} = k_y v_{De} = k_y \frac{\kappa T_e}{eB_o}. \tag{2.9.13}$$

v_{De} is the electron diamagnetic drift velocity defined in such a way that $-en_o v_{De}$ produces a diamagnetic current J_D [Eq. (1.8.13)] due to the electron pressure gradient. The wave given by the dispersion relation (2.9.12) is called the drift wave because the wave is carried by the diamagnetic drift of electrons (which is different from the guiding center drift introduced in Sect. 1.8). For a homogeneous plasma the drift wave frequency is zero. As can be seen from the derivation of Eq. (2.9.9), the drift wave is basically a vortex mode whose frequency is zero (or a purely damped mode in the presence of viscosity). It should further be noticed that if we compare the nonlinear term $\rho_s^2 \nabla_\perp^2 \phi$ with the linear term, $(T_e/e)\ell n(\omega_{ci}/n_o)$, in Eq. (2.9.9), the nonlinear term becomes comparable to the linear term if $k_\perp^2 \rho_s^2 e\phi/T_e \simeq \kappa\rho_s$. Since the inhomogeneous scale length κ^{-1} is much larger than the ion Larmor radius at the electron temperature, $\kappa\rho_s$ is ordinarily a very small quantity. This indicates that the drift wave becomes nonlinear even at a very small amplitude of the wave, $e\phi/T_e \simeq \kappa\rho_s$, if $k_\perp\rho_s \simeq 0(1)$. Because of this, in laboratory plasmas the drift waves are in fact observed almost always in the turbulent form. Hence, unlike other plasma waves, the proper description of the drift wave requires consideration of the nonlinear effect in Eq. (2.9.9). Equation (2.9.9) is called the Hasegawa-Mima equation.

2.10 Effect of Magnetic Field Curvature

When the magnetic line of force is curved, the dispersion relation of the Alfveń wave which propagates along the curved field line is modified (and can become unstable) because of the centrifugal force of plasma particles moving along the field line.

The Alfveń wave may be studied by constructing the vorticity equation from the MHD equation of motion. Instead of constructing the equation for $\Omega = \hat{\mathbf{b}} \cdot \nabla \times \mathbf{v}$, where $\hat{\mathbf{b}}$ is the unit vector in the direction of the magnetic field, we present an equivalent but alternative approach. We use the quasi-neutrality condition $\nabla \cdot \mathbf{J} = 0$ which may be written

$$\nabla \cdot \mathbf{J}_p + \nabla \cdot \mathbf{J}_D + \nabla \cdot \mathbf{J}_\parallel = 0. \tag{2.10.1}$$

Here

$$J_p = \hat{b} \times \frac{m_i n_o dv/dt}{B} \tag{2.10.2}$$

is the polarization current, Eq. (1.8.3),

$$J_D = \frac{\hat{b} \times \nabla p}{B} \tag{2.10.3}$$

is the diamagnetic current, Eq. (1.8.13), and $\mathbf{J}_{\parallel} = \hat{\mathbf{b}} J_{\parallel}$ is the parallel current. For simplicity, we consider the plasma density to be uniform (although the pressure will be nonuniform to maintain the equilibrium on the curved field line). We consider an incompressible wave ($\nabla \cdot \mathbf{v} = 0$) which propagates in the $\hat{\mathbf{b}}$ and $\hat{\mathbf{y}} (= \hat{\mathbf{b}} \times \hat{\mathbf{x}})$ directions with the magnetic field polarized in the inhomogeneous ($\hat{\mathbf{x}}$) direction (see Fig. 2.8). We take the Fourier amplitude expression for the local perturbation in the form

$$B_x(\ell, y, t) = B_{xk}(\ell) e^{i(k_y y - \omega t)} + c.c.,$$

where ℓ is the coordinate along the magnetic field.

Writing $\hat{\mathbf{b}} \cdot \nabla \times \mathbf{v} = \Omega$, the linear response of Eq. (2.10.2) gives

$$\nabla \cdot J_{p1} = -\frac{m_i n_o}{B_o} \frac{\partial \Omega}{\partial t}, \tag{2.10.4}$$

which is the time rate of change of the vorticity. If we use $\mathbf{E} \times \mathbf{B}$ drift for $\mathbf{v}$,

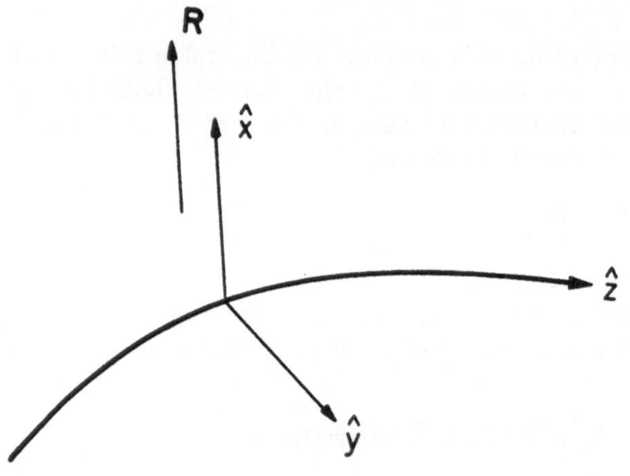

Fig. 2.8. Coordinate system in the curved field line

$$\Omega = -B_o \nabla \cdot (\frac{\mathbf{E}_1}{B_o^2}). \tag{2.10.5}$$

Hence the Fourier amplitude of $\nabla \cdot \mathbf{J}_{p1}$ becomes

$$\overline{\nabla \cdot \mathbf{J}_{p1}} = i\omega \frac{m_i n_o}{B_o} \Omega_k = \omega k_y \frac{m_i n_o}{B_o^2} E_{yk}, \tag{2.10.6}$$

where the bar on the left hand side indicates the Fourier transform. The linear response of $\mathbf{J}_D$ has two terms, one due to ∇p_1 and the other due to $\hat{\mathbf{b}}_1$. However $\hat{\mathbf{b}}_1$ does not contribute to the divergence of $\mathbf{J}_p$. This leads to

$$\overline{\nabla \cdot \mathbf{J}_{D1}} = ik_y p_k \hat{\mathbf{y}} \cdot \nabla \times \frac{\hat{\mathbf{b}}}{B_o}. \tag{2.10.7}$$

The pressure perturbation satisfies the incompressible equation of state, $dp_1/dt = 0$, i.e.,

$$p_k = \frac{1}{i\omega} \frac{1}{B_o} \frac{\partial p_o}{\partial x} E_{yk}, \tag{2.10.8}$$

where $\partial p_o / \partial x$ is the gradient of the background plasma pressure. Equations (2.10.7) and (2.10.8) give

$$\overline{\nabla \cdot \mathbf{J}_{D1}} = \frac{k_y}{\omega} \frac{1}{B_o} \frac{\partial p_o}{\partial x} \hat{\mathbf{y}} \cdot (\nabla \times \frac{\hat{\mathbf{b}}}{B_o}) E_{yk}. \tag{2.10.9}$$

Finally from Maxwell's equations,

$$J_{\parallel} = -\frac{ik_y}{\mu_o} B_{xk} = \frac{k_y}{\omega \mu_o} \frac{\partial E_{yk}}{\partial \ell} \tag{2.10.10}$$

while

$$\begin{aligned} \nabla \cdot \mathbf{J}_{\parallel} &= \nabla \cdot (\mathbf{J} \cdot \hat{\mathbf{b}}) \\ &= \mathbf{B}_o \cdot \nabla (\frac{J_{\parallel}}{B_o}). \end{aligned} \tag{2.10.11}$$

Hence

$$\overline{\nabla \cdot \mathbf{J}_{\parallel}} = \frac{k_y}{\omega \mu_o} B_o \frac{\partial}{\partial \ell} (\frac{1}{B_o} \frac{\partial E_{yk}}{\partial \ell}). \tag{2.10.12}$$

Substituting Eqs. (2.10.6), (2.10.9) and (2.10.12) into Eq. (2.10.1), we have,

$$B_o \frac{\partial}{\partial \ell}\left(\frac{1}{B_o}\frac{\partial E_{yk}}{\partial \ell}\right) + \frac{\omega^2 \mu_o m_i n_o}{B_o^2} E_{yk}$$

$$+ \frac{\mu_o}{B_o}\frac{\partial p_o}{\partial x}\hat{y}\cdot(\nabla \times \frac{\tilde{b}}{B_o})E_{yk} = 0. \qquad (2.10.13)$$

In this expression, we note that

$$\hat{\mathbf{y}}\cdot\nabla\times\frac{\mathbf{b}}{B_o} \equiv -\frac{R}{B_o},$$

where R is the x-component of the radius of curvature. Thus, Eq. (2.10.13) may be written,

$$B_o \frac{\partial}{\partial \ell}\left(\frac{1}{B_o}\frac{\partial E_{yk}}{\partial \ell}\right) + \frac{\omega^2}{v_A^2} E_{yk} - \frac{T}{m_i v_A^2 R^2}(\mathbf{R}\cdot\nabla \ell n p_o)E_{yk} = 0. \qquad (2.10.14a)$$

Equation (2.10.14a) is the desired local wave equation for an Alfveń wave on a curved field line. We note that the Alfveń wave dispersion relation given by the first two terms is modified due to the combination of the pressure gradient and the curvature. Since $T/m_i = v_{Ti}^2$, the last term can be identified as proportional to the centrifugal force. Since the plasma beta can be expressed as $\beta = 2v_{Ti}^2/v_A^2$, the local dispersion relation may be expressed in a WKB sense,

$$\frac{\omega^2}{v_A^2(\ell)} - k_\parallel^2(\ell) - \frac{\beta(\ell)}{2R^2}(\mathbf{R}\cdot\nabla \ell n p_o) = 0. \qquad (2.10.14b)$$

For example the ring current plasma in the magnetosphere has a pressure gradient directed toward the earth, while the radius of curvature is directed toward the sun, hence $\mathbf{R}\cdot\nabla \ell n p_o < 0$. We note that from Eq. (2.10.14b) the Alfveń frequency is reduced. If the plasma beta is larger than $2k_\parallel^2 |R/\nabla \ell n p_o|$, ω^2 becomes negative and the plasma can become unstable. This type of instability is called the ballooning mode because the field line in the large curvature and large pressure gradient region balloons out in the radial direction.

2.11 Wave Energy Density

In this section we introduce the concept of energy density of electromagnetic waves in a dispersive media where the dielectric constant is a function of the wave frequency and wave number as seen in Eq. (2.6.35).

For simplicity, let us first consider the case of a cold and unmagnetized plasma for which the plasma dielectric constant ϵ is a scalar and only a function of the wave frequency ω. The Fourier amplitude of the electric displacement $D(\omega)$ can then be expressed by

$$D(\omega) = \epsilon(\omega)E(\omega), \tag{2.11.1}$$

where $E(\omega)$ is the Fourier amplitude of the electric field. We are interested in deriving a proper expression for the wave energy density in this dielectric medium. We define the Fourier transform of D, ϵ, and E back to the real space,

$$\bar{D}(t) = \frac{1}{2\pi}\int_{-\infty}^{\infty} D(\omega)e^{-i\omega t}d\omega \tag{2.11.2}$$

$$\bar{\epsilon}(t) = \frac{1}{2\pi}\int_{-\infty}^{\infty} \epsilon(\omega)e^{-i\omega t}d\omega \tag{2.11.3}$$

$$\bar{E}(t) = \frac{1}{2\pi}\int_{-\infty}^{\infty} E(\omega)e^{-i\omega t}d\omega . \tag{2.11.4}$$

Then, $\bar{D}(t)$ satisfies, from the convolution formula,

$$\bar{D}(t) = \int_{-\infty}^{\infty} \bar{\epsilon}(t-t')\bar{E}(t')dt' . \tag{2.11.5}$$

In real space, Maxwell's equations read,

$$\frac{1}{\mu_0}\nabla\times\bar{\mathbf{B}} = \frac{\partial\bar{\mathbf{D}}}{\partial t} \tag{2.11.6}$$

and

$$\nabla\times\bar{\mathbf{E}} = -\frac{\partial\bar{\mathbf{B}}}{\partial t} . \tag{2.11.7}$$

The Poynting theorem can then be derived by constructing $\bar{\mathbf{E}}\cdot$(2.11.6) $-\bar{\mathbf{B}}\cdot$(2.11.7),

$$\overline{\mathbf{E}}\cdot\frac{\partial\overline{\mathbf{D}}}{\partial t}+\frac{1}{2\mu_o}\frac{\partial\overline{\mathbf{B}}^2}{\partial t}+\nabla\cdot(\overline{\mathbf{E}}\times\frac{\overline{\mathbf{B}}}{\mu_o})=0. \tag{2.11.8}$$

Thus, to construct the energy conservation equation, we should reduce $\overline{\mathbf{E}}\cdot\partial\overline{\mathbf{D}}/\partial t$ in the form of $\partial W_E/\partial t$, where W_E is the electric field energy of the wave. For this purpose we write $\overline{E}(t)$ as the product of an oscillating phase at frequency ω and its slowly varying amplitude,

$$\overline{E}(t)=\frac{1}{2}[E(\omega,t)e^{-i\omega t}+c.c.], \tag{2.11.9}$$

where $E(\omega, t)$ is in general a complex function also of ω, but $\overline{E}(t)$ is, of course, real. The electric displacement $\overline{D}(t)$ can then be written from Eq. (2.11.5),

$$\overline{D}(t)=\frac{1}{2}\int_{-\infty}^{\infty}\overline{\epsilon}(t-t')E(\omega,t')e^{-i\omega t'}dt'$$

$$+\ c.c.$$

$$=\frac{1}{2}e^{-i\omega t}\int_{-\infty}^{\infty}\overline{\epsilon}(\tau)E(\omega,t-\tau)e^{i\omega\tau}d\tau$$

$$+\ c.c. \tag{2.11.10}$$

The assumption of slowly varying amplitude allows us to expand $E(\omega, t-\tau)$ in powers of τ,

$$E(\omega,t-\tau)\simeq E(\omega,t)-\frac{\partial E}{\partial t}\tau. \tag{2.11.11}$$

If we then use the relations,

$$\int_{-\infty}^{\infty}\frac{\partial\overline{\epsilon}(\tau)}{\partial\tau}e^{i\omega\tau}d\tau=-i\omega\epsilon(\omega) \tag{2.11.12}$$

and

$$\int_{-\infty}^{\infty}\tau\overline{\epsilon}(\tau)e^{i\omega\tau}d\tau=-i\frac{\partial\epsilon}{\partial\omega}, \tag{2.11.13}$$

we have

$$\frac{\partial \overline{D}(t)}{\partial t} = -\frac{i\omega}{2}[\epsilon(\omega)E(\omega,t) + i\frac{\partial \epsilon}{\partial \omega}\frac{\partial E}{\partial t}]e^{-i\omega t}$$

$$+\epsilon\frac{\partial E}{\partial t}e^{-i\omega t} + \text{c.c.} \qquad (2.11.14)$$

If we now construct $\overline{E}\,\partial\overline{D}/\partial t$, and take a time average by neglecting the imaginary part of ϵ, we can finally obtain,

$$\overline{E}\frac{\partial \overline{D}}{\partial t} = \frac{\partial W_E}{\partial t},$$

$$\text{where } W_E = \frac{1}{2}\frac{\partial}{\partial \omega}(\omega\epsilon)\left[E(\omega,t)E^*(\omega,t)\right] \qquad (2.11.15)$$

is identified as the energy density of the wave electric field. The magnitude of the field energy $W_B = B^2/2\mu_o$ can also be expressed in terms of the electric field

$$W_B = \frac{1}{2\mu_o\omega^2}\left[k^2\mathbf{E}\cdot\mathbf{E}^* - (\mathbf{k}\cdot\mathbf{E}^*)(\mathbf{k}\cdot\mathbf{E})\right]. \qquad (2.11.16)$$

We can extend the derivation of the wave energy to a more general case in which the dielectric constant is a tensor. Then the wave energy density, W_w, can be expressed as a sum of W_E and W_B,

$$W_w = \frac{1}{2}\frac{\partial}{\partial \omega}\left\{\omega\mathbf{E}^*\cdot\overleftrightarrow{\epsilon'}\cdot\mathbf{E} - \frac{\epsilon_o c^2 k^2}{\omega}\mathbf{E}^*\cdot\mathbf{E} + \frac{\epsilon_o c^2}{\omega}(\mathbf{k}\cdot\mathbf{E}^*)(\mathbf{k}\cdot\mathbf{E})\right\}. \qquad (2.11.17)$$

where

$$\epsilon'_{\alpha\beta} = \frac{\epsilon_{\alpha\beta} + \epsilon^*_{\alpha\beta}}{2} \qquad (2.11.18)$$

is the Hermitian part of the dielectric tensor. When ϵ has a wavenumber dependency, the wave energy flux, $\mathbf{P}_w$, can also be defined in a similar manner by assuming a slowly varying electric field amplitude in space,

$$\mathbf{P}_w = -\frac{1}{2}\frac{\partial}{\partial \mathbf{k}}\left\{\omega\mathbf{E}^*\cdot\overleftrightarrow{\epsilon'}\cdot\mathbf{E} - \frac{\epsilon_o c^2 k^2}{\omega}\mathbf{E}^*\cdot\mathbf{E} + \frac{\epsilon_o c^2}{\omega}(\mathbf{k}\cdot\mathbf{E}^*)(\mathbf{k}\cdot\mathbf{E})\right\}. \qquad (2.11.19)$$

The energy conservation law now reads,

$$\frac{\partial W_w}{\partial t} + \nabla\cdot\mathbf{P}_w = -\,\mathrm{Im}(\mathbf{E}^*\cdot\overleftrightarrow{\epsilon}\cdot\mathbf{E}). \qquad (2.11.20)$$

2.12 Wave Kinetic Equation

In this section we derive the kinetic equation for the number density of a wave, I $(\mathbf{k}, \mathbf{x}, t)$ in an inhomogeneous plasma. The quantity I is defined as the ratio of the wave energy density to the central frequency of the wave,

$$I(\mathbf{k}, \mathbf{x}, t) = \frac{W_w(\mathbf{k}, \mathbf{x}, t)}{\omega(\mathbf{k}, \mathbf{x})}. \tag{2.12.1}$$

For the quantity I to have the proper dimension, the right hand side of this expression must be divided by the Planck constant h. However, when the wave is not quantized, the number density of the wave is commonly defined as the action of the wave as expressed in Eq. (2.12.1).

Let us first explain why the wave action may be defined as the ratio of the energy density to the frequency in accordance with Hamilton's action angle variables introduced in Section 1.1. For this purpose, we consider a harmonic oscillator which is represented by a particle motion in a parabolic potential well, $V = \omega_o^2 x^2/2$. The equation of motion of a particle with a unit mass in this potential well is given by

$$\left.\begin{aligned} &\frac{dv}{dt} = -\frac{\partial V}{\partial x} = -\omega_o^2 x \\ \text{and} \quad &\frac{dx}{dt} = v. \end{aligned}\right\} \tag{2.12.2}$$

The solution of these differential equations is readily obtained,

$$\left.\begin{aligned} x &= x_o \sin(\omega_o t + \alpha) \\ v &= \omega_o x_o \cos(\omega_o t + \alpha), \end{aligned}\right\} \tag{2.12.3}$$

where x_o and α are constants and represent the amplitude and phase of the oscillating motion. If we use a set of canonical variables $P = v$, $Q = x$, the Hamiltonian is given by

$$H = \frac{P^2}{2} + \frac{\omega_o^2 Q^2}{2}, \tag{2.12.4}$$

and Hamilton's equation of motion is given by the variation of the line integral of the Lagrangian,

$$\delta \int L dt = 0; \quad L = P\dot{Q} - H, \tag{2.12.5}$$

as

$$\frac{\partial H}{\partial P} = \dot{Q}, \quad \frac{\partial H}{\partial Q} = -\dot{P}. \tag{2.12.6}$$

Equations (2.12.4) and (2.12.6) are readily identified as equivalent to Eq. (2.12.2). Since this solution is periodic, the action J defined in Eq. (1.11.1) is given by

$$\begin{aligned} J &= \frac{1}{2\pi}\oint P dQ \\ &= \frac{1}{2\pi}\oint v^2 dt = \frac{1}{2\pi\omega_o}\int_0^{2\pi} \omega_o^2 x_o^2 \cos^2(\omega_o t + \alpha) d(\omega_o t) \\ &= \frac{\omega_o x_o^2}{2}. \end{aligned} \tag{2.12.7}$$

If we use the solution of Eq. (2.12.3) for x and v, H in Eq. (2.12.4) becomes $\omega_o^2 x_o^2/2$. Hence J in Eq. (2.12.7) is related to H through

$$J = H/\omega_o, \tag{2.12.8}$$

which is the ratio of energy to frequency. If we use the action-angle variables as the canonical momentum and coordinates, Hamilton's equation of motion becomes,

$$\frac{\partial H}{\partial J} = \dot{\theta} = \omega_o \tag{2.12.9}$$

and

$$\frac{\partial H}{\partial \theta} (= 0) = -\dot{J}. \tag{2.12.10}$$

Equation (2.12.9) corresponds to (2.12.8). Equation (2.12.10) represents the conservation of the action J.

Now the conservation of the number density of a wave I $(\mathbf{k}, \mathbf{x}, t)$ may be formally expressed in the $\mathbf{k} - \mathbf{x}$ phase space as

$$\frac{\partial I}{\partial t} + \dot{\mathbf{x}} \cdot \frac{\partial I}{\partial \mathbf{x}} + \dot{\mathbf{k}} \cdot \frac{\partial I}{\partial \mathbf{k}} = 0. \tag{2.12.11}$$

As in the case of the Vlasov equation, the quantities, $\dot{\mathbf{x}}$ and $\dot{\mathbf{k}}$ should be related to the field quantity which in this case is the property of the wave. For this purpose we consider the phase angle of the wave in a WKB sense for an inhomogeneous medium,

$$\theta = \int^{\mathbf{x}} \mathbf{k}\cdot d\mathbf{x} - \omega t \,. \tag{2.12.12}$$

The temporal variation of the phase angle is then given by

$$\dot{\theta} = \mathbf{k}\cdot\dot{\mathbf{x}} - \omega \,. \tag{2.12.13}$$

If we compare this expression with the Lagrangian L of Eq. (2.12.5) for a particle with Hamiltonian H (P, Q, t), momentum P(t), at coordinate position Q(t), we note a similarity if P is replaced by the wave number $\mathbf{k}$ and H(P, Q, t) by the local wave frequency $\omega(\mathbf{k}, \mathbf{x})$. We recall that Hamilton's equation of motion is obtained from the principle of least action, that is, from the variation of the line integral of the Lagrangian between two fixed points,

$$\delta \int_{t_2}^{t_2} L dt = 0 \,. \tag{2.12.14}$$

Thus we write $\dot{\theta}$ in Eq. (2.12.12) as a wave phase Lagrangian L_W

$$L_W (\equiv \dot{\theta}) = \mathbf{k}\cdot\dot{\mathbf{x}} - \omega(\mathbf{k}, \mathbf{x}) \tag{2.12.15}$$

and require minimization of the phase angle

$$\delta\theta = \delta \int_{t_1}^{t_2} L dt = 0 \,. \tag{2.12.16}$$

Hamilton's equations of motion are obtained by taking variations with respect to $\mathbf{k}$,

$$\dot{\mathbf{x}} = \frac{\partial \omega}{\partial \mathbf{k}} \tag{2.12.17}$$

and with respect to $\mathbf{x}$,

$$\dot{\mathbf{k}} = -\frac{\partial \omega}{\partial \mathbf{x}} \,. \tag{2.12.18}$$

Hence if the local dispersion relation is obtained for ω as a function of $\mathbf{k}$ and $\mathbf{x}$, $\dot{\mathbf{x}}$ and $\dot{\mathbf{k}}$ are derivable from Eqs. (2.12.17) and (2.12.18). If we substitute these expressions into Eq. (2.12.10), we obtain the wave kinetic equation,

$$\frac{\partial I}{\partial t} + \{I, \omega\} = 0. \tag{2.12.19}$$

where the Poisson bracket $\{I, \omega\}$ is given by

$$\{I, \omega\} = \frac{\partial \omega}{\partial \mathbf{k}} \cdot \frac{\partial I}{\partial \mathbf{x}} - \frac{\partial \omega}{\partial \mathbf{x}} \cdot \frac{\partial I}{\partial \mathbf{k}}. \tag{2.12.20}$$

It may be appropriate to discuss here the relation between the wave kinetic equation (2.12.19) and the wave energy conservation relation (2.11.20). If we use the definition of I in Eq. (2.12.1), Eq. (2.12.19) may be written using $\partial\omega/\partial t = 0$,

$$\frac{\partial W_w}{\partial t} + \frac{\partial \omega}{\partial \mathbf{k}} \cdot \frac{\partial W_w}{\partial \mathbf{x}} - \frac{\partial \omega}{\partial \mathbf{x}} \cdot \frac{\partial W_w}{\partial \mathbf{k}} = 0. \tag{2.12.21}$$

On the other hand, in the absence of dissipation, Eq. (2.11.21) may be formally reduced to

$$\frac{\partial W_w}{\partial t} + \frac{\partial w}{\partial \mathbf{k}} \cdot \frac{\partial W_w}{\partial \mathbf{x}} = 0. \tag{2.12.22}$$

Although Eq. (2.12.21) differs from Eq. (2.12.22), if one takes into account the dispersive effect which comes from the spatial dependence of the dielectric tensor $\overleftrightarrow{\epsilon}(\omega, \mathbf{k}, \mathbf{x})$, and retains the lowest order correction due to the nonlocal contribution of $\overleftrightarrow{\epsilon}$, one can derive the last term of Eq. (2.12.21) from the energy conservation equation (Kadomtsev, 1965). Hence the wave kinetic equation may be derived from the wave energy conservation equation under the adiabatic assumption, $\partial\omega/\partial t = 0$.

One may then wonder what is specific about the wave kinetic equation. The answer to this question exists in the process of the derivation of the wave kinetic equation presented here. We used only the conservation of wave action and did not assume a linear response of the plasma dielectric constant as we did for the wave energy conservation equation. This indicates that the wave kinetic equation is applicable even in a nonlinear case as long as wave action is conserved. Examples are the change of ω due to the ponderomotive force or the nonlinear wave-particle interactions such as the nonlinear

Landau damping (the Compton scattering) where $\partial\omega/\partial\mathbf{x}$ may be replaced by $(\partial\omega/\partial I)(\partial I/\partial\mathbf{x})$ based on the intensity dependent change of ω. In the case of a wave-wave interaction, the action of one wave is not conserved even if the total action may be conserved. This situation is analogous to the modification of the Vlasov equation due to two particle collisions.

References for Chapter 2

W. P. Allis, S. J. Buchsbaum and A. Bers, *Waves in Anisotropic Plasmas*, MIT Press, Cambridge 1963.

G. Bekefi, *Radiation Processes in Plasmas*, Wiley, New York 1966.

A. Hasegawa, *Plasma Instabilities and Nonlinear Effects*, Springer-Verlag, Heidelberg, 1975.

A. Hasegawa and C. Uberoi, *The Alfveń Wave*, U. S. DOE, Oak Ridge, 1985.

B. B. Kadomtsev, *Plasma Turbulence*, Academic Press, New York, 1965.

L. D. Landau and E. M. Lifshitz, *Electrodynamics of Continuous Media*, Addison-Wesley, Reading, 1959.

T. H. Stix, *The Theory of Plasma Waves*, McGraw-Hill, New York, 1962.

Chapter 3 Stationary Solar Plasma System

3.1 Introduction

In a strict sense a stationary state cannot exist in nature. Every phenomenon manifests itself as a transient in the ever-changing universe. In accordance with the expansion of the universe, new physical states and structures are continuously being born. Star formation is such an example. When the temperature of the expanding universe was still hot enough and above the ionization temperature of hydrogen atoms, the universe was in a plasma state consisting of electrons and protons. When the temperature decreased below that for ionization, electrons and protons recombined and produced neutral hydrogen atoms. Then the electric force, which had been the dominant force in the plasma-filled universe, was replaced by the gravitational force. The gravitational force governed the evolution of the universe thereafter. Clouds of hydrogen atoms formed here and there in the universe and eventually developed into protostars. The temperature and the density were drastically raised at the cores of protostars by the gravitational contraction. Thermonuclear reactions ultimately began and emitted photons. These photons ionized the surrounding neutral hydrogen and reproduced in places the plasma state in the universe. Our sun is a typical example of such stars. Solar atmospheric protons and electrons, created by photons emitted from the core, escape from the solar gravity and expand into interplanetary space. In their flow they interact with planets and other solid obstacles and finally encounter the interstellar gas.

The thermonuclear reaction in the sun is also a transient on the cosmic time scale. On the time scale we are interested in, however, the reaction can be considered a stationary phenomenon. Because of this fact the solar plasma system, which is driven by the thermonuclear reaction, can be considered as a stationary state.

We attempt here to describe the stationary features of the solar plasma system from a theoretical point of view. We try to do this in as self-contained a manner as possible based on known physical laws. Observational facts are used only for comparison with theoretical results or for making a proper choice of initial and boundary conditions.

In reality, the solar surface is not uniform, in both space and time; hence, neither is the solar plasma system. Interactions of solar plasma with other materials are therefore time-dependent. Most of the attractive and fascinating events in the solar plasma system come from its dynamic nature. The study of the stationary state, however, has its own value and also can provide a theoretical basis for revealing dynamic laws pertinent to space plasmas; this study is the main concern of Volume II.

3.2 Solar Plasma

The sun has a radius $R_\odot$ of approximately 7×10^5 km (110 times the earth radius), a mass $M_\odot$ of approximately 2×10^{30} kg (3.3×10^5 times the earth mass). It rotates approximately at a period of 27 days.

The core of the sun is fully ionized by its gravitational compression. The temperature and the density of the core plasma are roughly 2×10^7K and $10^{32}\,m^{-3}$, respectively. In addition, the plasma is ideally confined by gravity. The Lawson criterion that maintains the thermonuclear reaction in the core of the sun is thus more than sufficiently satisfied there.

The energy flux produced in the core by this thermonuclear reaction is radiatively transferred outwards. The resulting total light flux from the sun's surface amounts to 3.9×10^{26} joule/sec. Because of the high temperature of the core, 2×10^7K, the radiation ionizes the surrounding material (mostly hydrogen and a small amount of helium) as it propagates outwards.

The total radiation flux L through a spherical surface at a radius r is first absorbed by the fully ionized plasma and then, because of the denseness of the solar interior, reemitted as black body radiation. This process establishes radiative equilibrium.

If we let the net absorption coefficient of radiation per unit mass in the spherical shell with width dr at r be κ (opacity), then the temperature T (in degrees) at a radius r at radiative equilibrium is determined by

$$\frac{L}{4\pi r^2}\kappa mn = -\frac{d}{dr}(\sigma T^4), \tag{3.2.1}$$

where mn is the mass density and σ is the Stefan-Boltzmann constant (5.67×10^{-8} joule/m^{2}K^{4}sec). The temperature gradient is thus given by

$$\left.\frac{dT}{dr}\right|_{st} = -\frac{\kappa mn}{4\sigma T^3}\frac{L}{4\pi r^2}. \tag{3.2.2}$$

The subscript "st" stands for structure and identifies the temperature gradient due to radiative transfer.

As the radius increases, the temperature decreases and approaches the ionization temperature of hydrogen ($\sim 10^4 - 10^5$K). Near this point the incipient ionization contributes to a sharp decrease in the net absorption coefficient and, hence, a sharp drop of the temperature. The layer which includes this sharp temperature drop is called the "photosphere". In association with this sharp temperature gradient, an important dynamical process occurs: specifically, energy conversion from thermal energy to kinetic energy.

Mechanically, the pressure p in the solar interior is determined by the adiabatic law. Assuming the perfect gas law, we obtain the adiabatic temperature gradient as

$$\frac{dT}{dr}\Big|_{ad} = (1-\frac{1}{\gamma})\frac{T}{p}\frac{dp}{dr}, \qquad (3.2.3)$$

where γ is the adiabatic constant.

Furthermore, the pressure gradient force is balanced by the gravitational force. Thus,

$$\frac{dp}{dr} = -\frac{GM}{r^2}mn. \qquad (3.2.4)$$

From Eqs. (3.2.3) and (3.2.4), we obtain

$$\frac{dT}{dr}\Big|_{ad} = -(1-\frac{1}{\gamma})\frac{GMTmn}{r^2 p}, \qquad (3.2.5)$$

where G is the gravitation constant and M is the mass inside a sphere of radius r.

In the solar interior radiative transfer is so effective that there is no difference between the adiabatic gradient and the structural gradient. Therefore, no convection can occur. The temperature distribution is given by equating Eqs. (3.2.2) and (3.2.5):

$$4(1-\frac{1}{\gamma})\frac{\sigma T^4}{p} = \frac{\kappa L}{4\pi GM}. \qquad (3.2.6)$$

Near the sun's surface, however, where the structural temperature gradient becomes large, a difference will arise between the two gradients. As a result, a convection instability sets in. This unstable condition, called Schwarzschild's criterion, is expressed as

$$4(1-\frac{1}{\gamma})\frac{\sigma T^4}{p} < \frac{\kappa L}{4\pi GM}. \tag{3.2.7}$$

The region where convective motion sets in is called the "hydrogen convection zone". The thickness of this zone is about 15,000 km. In the photosphere, which is the top edge of the hydrogen convection zone and which can be as thin as 500 km, the temperature gradient is particularly enhanced because of the sharp increase of absorption. Consequently, the convection instability is expected to grow strongly there and to result in turbulent motion.

The temperature of the upper edge of the photosphere is reported to be about 4,500 K. Since no in situ energy source exists in the solar atmosphere, the temperature must decrease with increasing height, and thus no further ionization can occur. The sun is therefore reasonably expected to be surrounded by atmospheric gas in a cold diffusive equilibrium. Surprisingly, however, the temperature in the solar atmosphere starts increasing just above the photosphere. It rises to 10^6K at a height of 50,000 km and remains more or less constant above this height. The region where the temperature is about 10^6K is called the "corona" and the intermediate region between the photosphere and corona is called the "chromosphere". The structure of the solar temperature is schematically shown in Fig. 3.1.

A serious question thus arises: why and how is the solar atmosphere so strongly heated (from 4,000 K to 10^6K) despite there being no apparent in situ energy source. The lack of an obvious energy source and the fact that radiation is transparent suggest that the heating source might be waves associated with turbulent motion in the photosphere.

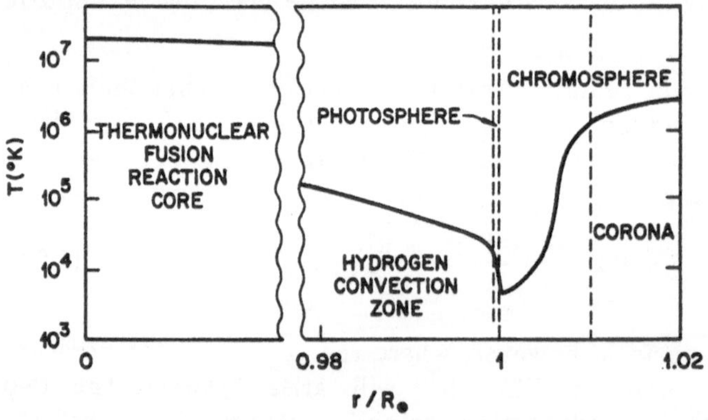

Fig. 3.1. Schematic profile of the solar plasma temperature

If neither heat sources nor heating processes are present in the solar atmosphere, the temperature profile above the photosphere will be governed by the heat conduction law (see Sect. 1.12),

$$\frac{1}{r^2}\frac{d}{dr}(r^2\kappa_T\frac{dT}{dr})=0, \tag{3.2.8}$$

where κ_T is the thermal conductivity given by Eq. (1.12.13);

$$\kappa_T=\kappa_{T0}T^{\frac{5}{2}}. \tag{3.2.9}$$

Substitution of Eq. (3.2.9) into Eq. (3.2.8) gives

$$T=T_0(\frac{r_0}{r})^{\frac{2}{7}}. \tag{3.2.10}$$

Obviously, the temperature decreases with height. Since the temperature of the upper edge of the photosphere is 4,500 K, the temperature of the solar atmosphere can never exceed the ionization temperature.

Therefore, the primary heat agent of the solar atmosphere must be mechanical (neutral) waves such as acoustic waves and gravity waves. Mechanical waves resulting from granulation in the photosphere might heat more efficiently in a lower density region. The temperature might eventually increase and reach the ionization temperature. Once plasma is generated, plasma waves can take part in further heating. The plasma waves can be those converted from mechanical (neutral) waves. Or, once plasma is generated, the solar magnetic field threading the photosphere can vibrate in association with turbulent motions in the lower photosphere where the medium is fully ionized. This magnetic field vibration can excite Alfveń waves or other magnetohydrodynamic waves (see Sect. 2.5) in the ionized chromosphere, which can further heat the plasma and may boost the temperature up to 10^6K. Quantitatively, no conclusive analysis or computer simulation has been done yet which can explain the solar chromospheric and coronal temperature structure. It is observationally certain, however, that the upper solar atmosphere consists of a high temperature ($\sim 10^6$K) plasma.

3.3 Solar Wind

In this section we study the radial behavior of the high temperature plasma (corona). The fundamental equations to describe the essence of the coronal behavior are the conservation equations discussed in Section 1.5, to which the effect of solar gravity is added. For simplicity, we assume that (i) no magnetic field exists, (ii) the plasma consists of electrons and protons only, and the behavior is time independent ($\partial/\partial t = 0$) and varies only radially.

In a spherical coordinate system, the particle, momentum and energy conservation equations for electrons and protons are

$$\frac{d}{dr} r^2 n_j v_j = 0 \tag{3.3.1}$$

$$n_j m_j v_j \frac{dv_j}{dr} = n_j e_j E - \frac{d}{dr} n_j T_j - \frac{n_j m_j G M_\odot}{r^2} \tag{3.3.2}$$

$$\frac{1}{r^2}\frac{d}{dr}[v_j r^2 (\frac{1}{2} n_j m_j v_j^2 + \frac{1}{\gamma - 1} n_j T_i)]$$

$$= -\frac{1}{r^2}\frac{d}{dr}(v_j r^2 n_j T_j) - \frac{v_j n_j m_j G M_\odot}{r^2} + \frac{1}{r^2}\frac{d}{dr}(r^2 \kappa_T \frac{dT_j}{dr}), \tag{3.3.3}$$

where j represents either electrons (e) or ions (i) and $M_\odot$ is the solar mass. In the stationary state, $n_e \cong n_i (\equiv n)$ and $v_e \cong v_i (\equiv v)$; otherwise, net charge and current sources must exist in the entire solar system, which is unrealistic. Actually, an electric field will force electrons to move together with protons. Such an electric field is called an "ambipolar" electric field.

Elimination of the electric field in Eq. (3.3.2) yields

$$m v \frac{dv}{dr} = -\frac{1}{n}\frac{d}{dr} n T - \frac{G M_\odot m}{r^2}, \tag{3.3.4}$$

where $m = m_e + m_i$ and $T = T_e + T_i$. Equation (3.3.1) gives

$$n v r^2 = \text{const.} \tag{3.3.5}$$

Substitution of Eq. (3.3.5) into Eq. (3.3.4) yields

$$\left(mv - \frac{T}{v}\right)\frac{dv}{dr} = -r^2\frac{d}{dr}\left(\frac{T}{r^2}\right) - \frac{GM_\odot m}{r^2}. \tag{3.3.6}$$

This equation is rewritten in the following normalized form:

$$\left(1 - \frac{\overline{T}}{\overline{K}}\right)\frac{d\overline{K}}{d\xi} = -2\xi^2\frac{d}{d\xi}\left(\frac{\overline{T}}{\xi^2}\right) - \frac{2\overline{G}}{\xi^2} \tag{3.3.7}$$

with

$$\xi = r/r_o \tag{3.3.8}$$

$$\overline{K} = (v/v_T)^2 \tag{3.3.9}$$

$$\overline{T} = T/T_o \tag{3.3.10}$$

$$\overline{G} = (v_{gr}/v_T)^2\,, \tag{3.3.11}$$

where $v_T[=(T_o/m)^{1/2}]$ is the thermal velocity and $v_{gr}[=(GM_\odot/r_o)^{1/2}]$ is the escape velocity; the subscript "o" indicates the corresponding value at a reference radius $r = r_o$. Equation (3.3.7) cannot be solved by itself. An additional equation relating the temperature change to the density and/or velocity change, i.e., Eq. (3.3.3), must be added for closure.

Let us consider here two cases where the solar atmospheric plasma expands (i) isothermally and (ii) adiabatically.

Isothermal Case

Deep in the corona the temperature is sustained by heating from below. Therefore, the isothermal assumption is expected to apply. In this case we can put $\overline{T} = 1$ in Eq. (3.3.10). Then, Eq. (3.3.7) reduces to

$$\left(1 - \frac{1}{\overline{K}}\right)\frac{d\overline{K}}{d\xi} = \frac{2}{\xi}\left(2 - \frac{\overline{G}}{\xi}\right). \tag{3.3.12}$$

Formally, this equation is solved to give

$$\overline{K} - \ln\overline{K} = 4\ln\xi + \frac{2\overline{G}}{\xi} + C\,, \tag{3.3.13}$$

where C is an integration constant determined by a boundary condition somewhere, possibly at $\xi = 1$ or at $\xi = \infty$.

In order to see the physical properties of the solution, Eq. (3.3.12) is more instructive than Eq. (3.3.13). Let us take a reference radius $r = r_0$ ($\xi = 1$) where the gravitational force exceeds the expansion force due to pressure and where the plasma flow is below the thermal velocity [i.e., $v_{gr} > \sqrt{2} v_T$ (or $\overline{G}/2 > 1$) and $\overline{K}(\xi = 1) < 1$)]. This choice will probably be in the lower part of the corona. Then, from Eq. (3.3.12) we have

$$\frac{d\overline{K}}{d\xi} > 0 . \tag{3.3.14}$$

This indicates that the solar plasma is radially accelerated. When the accelerated plasma reaches a radius where $\xi = \overline{G}/2$ (>1), the right hand side of Eq. (3.3.12) vanishes, indicating termination of the acceleration. If the flow speed has not reached the thermal speed v_T, then the flow begins to decelerate continuously. Thus, the flow will stop at some long distance. This solution is the so-called "solar breeze" solution, as is schematically shown in Fig. 3.2.

From Eq. (3.3.13) the distant solution falls off as

$$\overline{K} \approx \xi^{-4} . \tag{3.3.15}$$

Using this relation with Eq. (3.3.5), we have

$$n = \text{const.} (\neq 0) \quad \text{at } \xi = \infty . \tag{3.3.16}$$

This solution seems unrealistic unless sufficient plasma sources exist at great distances, since the total plasma content in a unit shell will

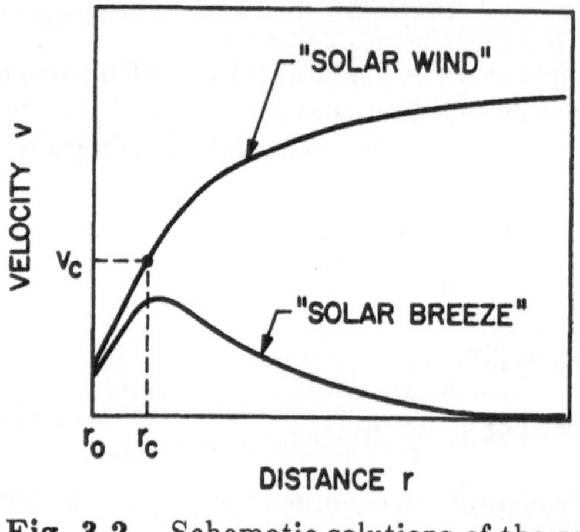

Fig. 3.2. Schematic solutions of the solar wind and solar breeze

diverge with distance ξ. Therefore, we shall try to find another solution by going back to Eq. (3.3.12).

In obtaining the previous solutions, we assumed that at $\xi = \overline{G}/2$ the flow speed had not reached the thermal speed ($\overline{K} < 1$). Let us abandon this assumption and assume that $\overline{K} = 1$ at $\xi = \overline{G}/2$. This condition satisfies Eq. (3.3.12). In addition, $d\overline{K}/d\xi$ is still positive at this position. Thus, the flow becomes supersonic at $\xi = \overline{G}/2 (\equiv \xi_c)$ and continues to accelerate beyond this position. This is the trans-sonic solution which is more familiarly called the "solar wind". The solar wind solution is also schematically shown in Fig. 3.2.

In the solar wind solution, the distant solution is approximated from Eq. (3.3.13), by

$$\overline{K} \approx 4 \ell n \xi . \tag{3.3.17}$$

Substituting Eq. (3.3.17) into Eq. (3.3.5), gives the distant density

$$n \propto \xi^{-2} (\ell n \xi)^{-1/2} . \tag{3.3.18}$$

In this solution the total plasma content N in a unit shell obeys $N \propto (\ell n \xi)^{-1/2}$ which is a decreasing function of distance ξ. This indicates that the supersonic solution is a realizable solution in the solar system even if no *ad hoc* particle source is assumed outside the solar system.

Adiabatic Case

The adiabatic law, $pn^{-\gamma} = \text{const}$, is rewritten in terms of T and n as

$$Tn^{-(\gamma-1)} = \text{const}. \tag{3.3.19}$$

From this relation we obtain

$$\frac{1}{n}\frac{d}{dr} nT = \frac{\gamma}{\gamma-1}\frac{dT}{dr} . \tag{3.3.20}$$

Substitution of Eq. (3.3.20) into Eq. (3.3.4) yields

$$\frac{1}{2} mv^2 + \frac{\gamma}{\gamma-1} T - GM_\odot \frac{m}{r} = \text{const}. \tag{3.3.21}$$

Alternatively, this relation can be obtained from Eq. (3.3.3) by neglecting thermal conduction and using Eq. (3.3.5). The normalized form of Eq. (3.3.21) is given by

$$\frac{1}{2}\overline{K}+\frac{\gamma}{\gamma-1}\overline{T}-\frac{\overline{G}}{\xi}=E_o\,, \tag{3.3.22}$$

where E_o is an integration constant given by

$$E_o=\frac{1}{2}\overline{K}_o+\frac{\gamma}{\gamma-1}-\overline{G} \tag{3.3.23}$$

and $\overline{K}_o$ is the value of $\overline{K}$ at $\xi=1$.

Differentiation of Eq. (3.3.22) with respect to ξ yields

$$\frac{d\overline{T}}{d\xi}=-\frac{\gamma-1}{\gamma}\left(\frac{1}{2}\frac{d\overline{K}}{d\xi}+\frac{\overline{G}}{\xi^2}\right). \tag{3.3.24}$$

Using this relation in Eq. (3.3.7) obtains

$$\left(1-\frac{\gamma\overline{T}}{\overline{K}}\right)\frac{d\overline{K}}{d\xi}=\frac{2}{\xi}\left(2\gamma\overline{T}-\frac{\overline{G}}{\xi}\right). \tag{3.3.25}$$

This equation indicates that discussions analogous to the ones for the isothermal case can be applied, even though the behavior is nonlinear in the sense that the temperature change couples with the flow change.

The trans-sonic solution must satisfy following relations

$$\begin{aligned} \overline{K}_c &= \gamma\overline{T}_c \\ \xi_c &= \frac{\overline{G}}{2\gamma\overline{T}_c} \\ \overline{T}_c &= \frac{2(\gamma-1)E_o}{\gamma(5-3\gamma)}\,. \end{aligned} \tag{3.3.26}$$

In order for a trans-sonic solution to be realized, there must exist a region in the corona where the expansion (thermal) energy of the plasma is not too large to allow free expansion but not too small to prohibit expansion. This condition is given, from Eqs. (3.3.23) and (3.3.25), by

$$\frac{\gamma}{\gamma-1}>\overline{G}>2\gamma. \tag{3.3.27}$$

For example, let us take $r_o=R_\odot$ (solar radius) and $\overline{G}=10$. Then from Eq. (3.3.11) we have

$$T \simeq 2.3 \times 10^6 K,$$

using $G = 6.67 \times 10^{-11}$ $m^3/kg\ sec^2$, $M_\odot = 1.99 \times 10^{30}$ kg, $m = 1.67 \times 10^{-27}$ kg, $R_\odot = 6.96 \times 10^5$ km and the Boltzmann constant $= 1.38 \times 10^{-23}$ joule/deg. This is a very familiar temperature. Inversely speaking, when the corona is heated up to 10^6K, the solar wind starts expanding into interplanetary space.

Equation (3.3.21) gives the total energy flux carried away from the sun by the solar wind as

$$F_o = 4\pi r^2 nv\left(\frac{1}{2}mv^2 + \frac{\gamma}{\gamma-1}T - \frac{GM_\odot m}{r}\right). \tag{3.3.28}$$

Satellite observations tell us that this flux amounts to $10^{20\sim 21}$ joule/sec, which is much smaller than the radiation flux of 4×10^{26} joule/sec.

We note that the relation of $\overline{T}_c$ in Eq. (3.3.26) implies that when $\gamma = 5/3$, the trans-sonic solution does not exist. In reality, however, other terms neglected in our fundamental equations such as the thermal conduction, the solar magnetic field and the multiplicity of particle species can participate in the evolution and modify the result.

Energy Transfer in Solar Wind

The one-fluid approximation is appropriate for describing the overall behavior near the sun where the plasma density is rather high ($n \approx 10^{20} m^{-3}$) and collisions between electrons and protons are frequent enough to maintain equipartition of the thermal energy. As the solar wind supersonically expands away from the sun, however, electrons and protons tend to expand independently because of their infrequent collisions with one another. Furthermore, the solar (interplanetary) magnetic field, which is stretched out by the expanding plasma, influences the behaviors of electrons and protons.

Since the subtleties of these influences are beyond the scope of this book, we here only briefly describe the expansion and termination of the supersonic solar plasma (solar wind) at large distances.

The density and the solar wind speed at large distances may be well represented by the solutions of a one-fluid model; that is, the density varies nearly inversely as r^2 and the speed is nearly constant. The temperature, however, behaves in a more complicated way, since the energy transport of electrons and protons differs due to the difference in collision frequencies and due to the presence of the interplanetary magnetic field (see Sect. 1.12).

The magnetic field stretched out from the sun's surface expands radially with the solar wind. Since the field lines are fixed to the sun's rotating frame, (see Sect. 1.6), the field lines form conical Archimedean spirals at large distances. This can be seen by considering the distant plasma motion in the sun's spherical frame (r, θ, ϕ). In this frame,

$$v_r = v_s$$

$$v_\theta = 0 \tag{3.3.29}$$

$$v_\phi \simeq - r\Omega \sin\theta ,$$

where v_s is the solar wind velocity and Ω is the angular velocity of the sun. The distant spirals are therefore given by

$$\frac{1}{r}\frac{dr}{d\phi} = - \frac{v_s}{r\Omega \sin\theta} . \tag{3.3.30}$$

The distant magnetic field that satisfies $\nabla \cdot \mathbf{B} = 0$ can then be given by

$$B_r = B_o\left(\frac{r_o}{r}\right)^2$$

$$B_\theta = 0 \tag{3.3.31}$$

$$B_\phi = - \frac{B_o r_o^2 \Omega \sin\theta}{r v_s} ,$$

where B_o is the field strength at some reference level $r_o(\cong R_\odot)$. This relation indicates that the magnetic field strength at large distances varies inversely as r in the ecliptic plane ($\theta = \pi/2$) and as r^2 in the polar region ($\theta = 0, \pi$). The direction of the field lines is therefore almost azimuthal in the ecliptic plane and radial in the polar region. The spiral structure of the interplanetary magnetic field in the ecliptic plane is shown in Fig. 3.3.

An important effect of the magnetic field on the temperature structure in the solar wind is to force particles to gyrate tightly around it. In particular, the thermal conductivity in the direction perpendicular to the magnetic field is reduced by a factor of $(\omega_c \tau)^2$ compared to that parallel to, or in the absence of, the field (see Sect. 1.12), where ω_c is the cyclotron frequency and τ is the collision time. The magnetic insulation of thermal conduction is therefore expected to take place near the ecliptic plane at large distances where

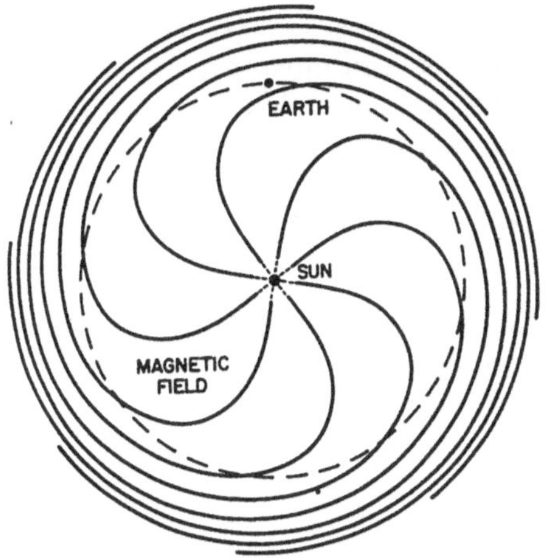

Fig. 3.3. Spiral structure of the interplanetary magnetic field (IMF) in the equatorial plane

the spiral direction is almost azimuthal. Elsewhere the thermal conduction is governed by collisions between particles.

At large distances the solar gravity is negligible, and the solar wind speed and density are roughly constant and inversely proportional to r^2, from Eqs. (3.3.17) and (3.3.18) respectively. When the thermal conduction is negligible, the temperature is governed by adiabatic expansion. Since $T \propto n^{\gamma-1}$ from Eq. (3.3.19) and $n \propto r^{-2}$, we obtain

$$T_i \propto r^{-2(\gamma-1)} = r^{-\frac{4}{3}} \left(\gamma = \frac{5}{3}\right). \tag{3.3.32}$$

In the opposite case, the temperature is governed by heat conduction; from Eq. (3.2.10) we have

$$T_i \propto r^{-\frac{2}{7}}. \tag{3.3.33}$$

Summarizing the above discussions, we can generally say that the temperature of electrons (or protons) falls off as $r^{-2/7}$ at intermediate distances where the interplanetary magnetic field has a substantial radial component, while as $r^{-4/3}$ at large distances where the field lines are nearly azimuthal and the radial thermal conduction is inhibited.

Another important issue regarding the temperature is anisotropy with respect to the direction of the magnetic field. If the collision time is smaller than the characteristic expansion time, i.e., $[v_s d(\ell n n)/dr]^{-1}$, then particles are scattered in all directions and anisotropy is reduced, even for particles tightly bound by the magnetic field. Electrons are said to be collisional in this sense, so that the above summary statement may be applied. Protons, on the other hand, are considered to be collisionless. Therefore, anisotropy is expected to arise in the proton temperature.

In the above discussions, only Coulomb collisions are considered. When the plasma distributions, e.g., of the velocity and the temperature, deviate from equilibrium, plasma instabilities are excited as will be seen in Volume II, and consequently the deviations will be reduced.

Formation of Heliosphere

Now a question may be raised: What is the destination of the solar wind?

The space in our galaxy is not empty in a rigorous sense, but is filled with interstellar gas, the interstellar magnetic field and cosmic rays. These quantities exert forces on the solar wind. Since the energy flux of the solar wind per unit area decreases in proportion to r^{-2}, there must exist a transition region where the forces between the solar wind and the interstellar medium balance. The solar wind velocity is almost constant at great distances and remains supersonic. The transition is therefore expected to form a shock structure and the supersonic solar wind expansion will terminate at a finite distance. The supersonic expansion is confined by the interstellar medium within a cavity called the "heliosphere".

The shape of the heliosphere cannot be drawn without sufficient information about the interstellar medium: whether the interstellar medium is a plasma or a neutral gas, what its speed is with respect to the sun, how strong the interstellar magnetic field is, in which direction it points, what its spatial structure is, etc.

Low frequency radio wave absorption and the frequency dispersion of pulsar emissions indicate that there certainly exists ionized interstellar gas with a density of $\sim 10^5 m^{-3}$. Measurements of the Zeeman splitting effect and the Faraday rotation of radiation from a pulsar suggest that there exists an interstellar magnetic field of the order of $\sim 10^{-9}$ tesla. The 21 cm emission and Lyman α absorption measurements also suggest that there exists a neutral interstellar gas with a density of $\sim 10^5 m^{-3}$. Since this observational information is still limited, the specific heliospheric structure remains uncertain.

The overall structure of the heliosphere, however, can be estimated by the pressure balance relation,

$$\frac{1}{2}m_s n_s v_s^2 + \frac{B_s^2}{2\mu_o} = \frac{1}{2}m_g n_g v_g^2 + n_g T_g + \frac{B_g^2}{2\mu_o}, \qquad (3.3.34)$$

where subscript s stands for the solar wind, subscript g for the interstellar plasma and μ_o is the permeability. Since $n_s \propto r^{-2}$, $v_s \approx$ constant and $B_s \propto r^{-1}$ near the ecliptic plane, the Alfveǹ Mach number

$$M_A = \left(\frac{\mu_o n_s m_s v_s^2}{B_s^2} \right)^{\frac{1}{2}} \qquad (3.3.35)$$

tends to a constant value ($M_A \approx 10$) at large distances. This indicates that the interplanetary magnetic pressure is negligible compared with the solar wind dynamic pressure. From Eq. (3.3.34) we obtain the distance of the shock front r_h as

$$r_h \approx 80 \text{ AU}, \qquad (3.3.36)$$

where AU is an astronomical unit, that is, the distance between the sun and the earth (1.5×10^8km). In obtaining Eq. (3.3.36) we assumed $m_s = m_g = 1.6 \times 10^{-27}$kg, $n_g = 10^5 m^{-3}$, $T_g = 10^4 K = 1.38 \times 10^{-19}$ joule, $v_g = 20$ km sec^{-1} (relative speed to the sun), $B_g = 5 \times 10^{-10}$ tesla and $n_s = n_{se}(r_e/r_h)^2$ with $n_{se} = 5 \times 10^6 m^{-3}$.

Beyond the shock front the solar wind blows subsonically and interacts more delicately with the interstellar medium. The subsonic flow will eventually be blown away in the direction opposite to the sun's movement in the interstellar gas (comet-like structure). The presence of the interstellar magnetic field can also macroscopically deform the subsonic flow structure.

In the above discussion the role of the neutral interstellar gas is also ignored. The photoionization of the neutral gas by solar radiation and charge exchange with the solar wind plasma can provide additional particle, momentum and energy sources. Thus, the fundamental equations, i.e., Eqs. (3.3.1) through (3.3.3), must be supplemented with these sources on the right hand side. However, since this problem is beyond the scope of this book, we wiii not discuss it here.

3.4 Earth's Magnetosphere

In the previous section we have seen that a supersonic plasma flow called the solar wind can steadily and radially blow in interplanetary space at a distance beyond the trans-sonic point ($r \approx 5\ R_\odot$) as long as the plasma conditions in the corona are stationary and isotropic. Solar wind conditions may change from time to time depending on changes near the solar surface. However, since this volume is primarily concerned with a stationary plasma state, we will treat the solar wind as a stationary wind.

Inside the heliosphere there are many solid obstacles such as planets and their satellites. Based on our knowledge from space craft movements, the solar wind plasma is extremely tenuous. Viewed from the size of a planet, however, it is so dense that the solar wind particles cannot pass by without experiencing collisions, since the interparticle distance is as small as 10^{-2}m. Thus, the solar wind can strongly interact with a planet. On the other hand, on the heliospheric scale, the sizes of the planets are too small to change the overall solar wind structure. Furthermore, the distance between any two planets is extremely large compared to their sizes. These facts allow us to regard the solar wind as a given input and treat the interaction with each planet independently.

In this section we focus on the interaction between the solar wind and our earth and stationary plasma structures resulting from the interaction of the earth's atmosphere.

The atmosphere on the earth is filled with a neutral gas. As we will see in Section 3.5, the atmospheric density decreases exponentially with height. The scale height is $5 \sim 10$ km near the earth and the density at the earth's surface is of the order of $10^{25}\mathrm{m}^{-3}$. Solar radiation photoionizes the atmosphere above a height of about 100 km and creates the ionosphere. The maximum plasma density occurs at an altitude of about 250 km ($\sim 10^{12}\mathrm{m}^{-3}$) where the neutral gas density is still much higher ($\sim 10^{15}\mathrm{m}^{-3}$) than the plasma density (see Fig. 3.20). The plasma density then decreases with height. At about 3,000 km the atmospheric gas becomes fully ionized with a density of $\sim 10^{9\sim 10}\mathrm{m}^{-3}$. This photoionized plasma and the neutral atmospheric gas could directly interact with the solar wind plasma if they were directly exposed to the solar wind. In reality, however, we must consider the effect of the earth's magnetic field, also called the geomagnetic field, or the terrestrial magnetic field. As in the discussion of the heliosphere in the last part of Section 3.3, the geomagnetic field participates importantly in the interaction between the solar wind and the earth.

In the geomagnetic spherical coordinate system (r, θ, ϕ) where the geomagnetic dipole moment is oriented in the negative direction of the z axis, i.e., $\theta=\pi$, the field vector is given by

$$\mathbf{B}=\nabla\left(\frac{M\cos\theta}{r^2}\right), \tag{3.4.1}$$

where M is the magnitude of the dipole moment ($M=8\times10^{15}$ tesla m^3).

The solar wind pressure at the earth's orbit is about 5×10^{-10} newton/m^2. The magnetic pressure on the earth's surface is about 4×10^{-4} newton/m^2, much larger than the solar wind pressure. Since the dipole field pressure decreases as r^{-6}, the geomagnetic field pressure becomes equal to the solar wind pressure at a distance of approximately $r=10\,R_E$ where R_E is the earth's radius $\approx$ 6370 km. The atmospheric plasma pressure, on the other hand, is at most 5×10^{-10} newton/m^2 at the height of 3000 km, i.e., at $r\approx1.5R_E$, where the temperature is assumed to be 5000 K and the density is $10^{10}m^{-3}$. Since the atmospheric plasma pressure becomes negligibly small at such a great distance as $10R_E$, it is expected that the geomagnetic field is the first barrier to stand in the way of the solar wind. If so, the barrier location must be at $r\sim10R_E$, where the solar wind and geomagnetic pressures become equal.

There is another force of earth origin that can interact with the solar wind. It is the centrifugal force due to the corotation of the atmospheric gas with the earth's rotation. The corotation pressure, p_c, is given by

$$p_c=nm(r\Omega_E)^2, \tag{3.4.2}$$

where r is the radial distance and Ω_E is the angular speed of the earth's rotation ($\Omega_E=7.27\times10^{-5}\,sec^{-1}$). Let us evaluate how much density is required to balance with the solar wind pressure at $r=10R_E$. Since the solar wind pressure is about 5×10^{-10} newton/m^2, the density required is calculated to be $n=1.46\times10^{10}m^{-3}$. This density is unrealistically large. Thus, the centrifugal force does not contribute to the formation of the magnetopause in the case of the earth. As will be discussed later in this section, however, corotation plays an important role in the internal structure of the magnetosphere.

The above argument is based on a magnetohydrodynamic (MHD) approximation, which is valid only for spatial scales much larger than the Larmor radius of a charged particle in the solar wind (see Sect. 1.6). From Eq. (1.4.9) we can estimate the Larmor radii of solar wind electrons and protons with respect to the geomagnetic field at

$r = 10R_E$,

$$\rho_e \approx 200 \text{ m} \quad \text{and} \quad \rho_p \approx 130 \text{ km},$$

where the proton velocity perpendicular to the geomagnetic field is taken to be the solar wind velocity of 400 km/sec (the proton thermal velocity is about 20 km/sec) and the electron perpendicular velocity is given by a thermal velocity corresponding to a temperature of 10^5K; the geomagnetic field intensity is taken to be 3×10^{-8} tesla. Thus, both the electron and proton Larmor radii are small enough to satisfy the necessary condition for using the MHD description for an overall discussion. The sufficient condition is the establishment of Maxwellian distribution, which is difficult to prove a priori. However, we here assume that the Maxwellian is maintained.

The above simple discussion suggests that the solar wind would be locally blocked by the geomagnetic field pressure at a distance of about 10 R_E from the earth. In addition, gas dynamics tells us that when a supersonic flow collides with a solid obstacle, a standing shock is created on the upstream side of the obstacle. The width of the shock can be of the order of the mean free path in the flow, since the supersonic flow energy is randomized by collisions and converted into thermal energy. This knowledge from gas dynamics suggests to us that a shock will be formed on the solar side of the earth where the solar wind collides against the geomagnetic field.

Plasma physics, on the other hand, tells us that the mean free path of electrons, λ_e, in a fully ionized gas is given by $\lambda_e = v_T/\nu_e$. Using Eq. (1.4.14), we have

$$\lambda_e = 1.5\times10^{17}\frac{T_e^2}{n\ell n\lambda}. \tag{3.4.3}$$

Substituting $n = 5\times10^6\text{m}^{-3}$, $T_e = 10$ eV ($\sim 10^5$K) and $\ell n\lambda = 20$, we obtain $\lambda_e \approx 1.5\times10^8$ km $\approx$1AU. This leads us to a dilemma: the fluid description we have used in this chapter may not be valid. As discussed in Section 1.6, however, we can avoid this dilemma by optimistically assuming that global particle distributions are always maintained as Maxwellian by some unknown processes.

In the treatment of the solar wind-earth interaction as well, we rely on this optimism. Specifically, when a solar wind particle encounters the geomagnetic field, it does not behave as a single particle but as one member of the group. If this were not the case, the geomagnetic field would remain unaffected by the solar wind even if the geomagnetic field pressure were smaller than the solar wind dynamic pressure. This is unrealistic, however, since when the Larmor radius is small compared with the characteristic dimension of interest, a diamagnetic current will be generated and the magnetic field will be deformed, as discussed in Sections 1.8 and 1.10. This implies that our optimism is not unrealistic.

In an ordinary hydrodynamic shock the force to decelerate a supersonic flow is the increased gas pressure in front of the obstacle. If this increase were due solely to an adiabatic compression, no shock would occur. A shock can be created because the temperature is more than adiabatically enhanced due to thermalization of the directed flow energy by collisions. In other words, because of the appearance of this additional stopping force over the mean free path, a shock is created and its width becomes of the order of the mean free path.

Extension of this physical consideration to the present case where the obstacle is the geomagnetic field reminds us that not only the pressure force but also the Ampere force can act to decelerate the solar wind. The origin of the Ampere force is the diamagnetic current generated at the expense of the distant geomagnetic energy which is swept towards the earth by the solar wind. The characteristic scale length of the guiding center currents is the Larmor radius. Since the plasma energy is considered to dominate the magnetic energy near the shock front, the characteristic length may be given by the proton Larmor radius. Hence, a shock can be created and the width of the shock transition layer can be of the order of the proton Larmor radius ($\rho_p \approx 100$ km), which is much smaller than the geomagnetic scale. Such a shock is called a "collisionless shock".

Downstream of such a shock front, the flow is subsonic. The subsonic flow will eventually be completely blocked by the squashed geomagnetic field and go around it just like a flow passing around a blunt body. Such a magnetic barrier impeding the intrusion of the solar wind is called a "magnetosphere".

Since the magnetohydrodynamic discontinuous structure is a common phenomenon of solar wind plasma physics (e.g., interplanetary shocks, shocks in front of planets, shocks with interstellar medium) we shall survey its general features in the following subsection.

Magnetohydrodynamic Discontinuities

In order to treat magnetohydrodynamic discontinuities, let us here reproduce the magnetohydrodynamic equations derived in Section 1.6 in the conservation form (after eliminating the electric field):

$$\frac{\partial}{\partial t}nm = -\nabla\cdot nm\mathbf{v} \tag{3.4.4}$$

$$\frac{\partial}{\partial t}nm\mathbf{v} = -\nabla\cdot(nm\mathbf{v}\mathbf{v} - \frac{\mathbf{B}\mathbf{B}}{\mu_o}) - \nabla(p+\frac{B^2}{2\mu_o}) \tag{3.4.5}$$

$$\frac{\partial U}{\partial t} = -\nabla\cdot\{(U+p+\frac{B^2}{2\mu_o})\mathbf{v} - \frac{\mathbf{v}\cdot\mathbf{B}}{\mu_o}\mathbf{B}\} \tag{3.4.6}$$

$$\frac{\partial \mathbf{B}}{\partial t} = \nabla\times(\mathbf{v}\times\mathbf{B}) \tag{3.4.7}$$

$$\nabla\cdot\mathbf{B} = 0 \tag{3.4.8}$$

where

$$U = \frac{1}{2}nmv^2 + \frac{1}{\gamma-1}p + \frac{B^2}{2\mu_o}. \tag{3.4.9}$$

From Eqs. (3.4.4), (3.4.5) and (3.4.6), the following jump conditions across an infinitesimally thin transition layer are derived

$$[nmv_n] = 0 \tag{3.4.10}$$

$$[nmv_n\mathbf{v}_t - \frac{B_n\mathbf{B}_t}{\mu_o}] = 0 \tag{3.4.11}$$

$$[nmv_n^2 + p + \frac{B_t^2-B_n^2}{2\mu_o}] = 0 \tag{3.4.12}$$

$$[\{\frac{1}{2}nm(v_n^2+v_t^2) + \frac{\gamma}{\gamma-1}p + \frac{B_t^2}{\mu_o}\}v_n - \frac{B_nB_t}{\mu_o}v_t] = 0, \tag{3.4.13}$$

where $[f] = f_2 - f_1$, the subscripts 1 and 2 representing the values of f on the upstream and downstream sides of the transition layer, respectively; the subscripts n and t represent components normal and tangential to the transition layer, respectively. These conditions are called "Rankine-Hugoniot" conditions.

We have two more independent equations, which are derived from Eqs. (3.4.8) and (3.4.7), namely,

$$[B_n] = 0 \tag{3.4.14}$$

$$[v_n \mathbf{B}_t - B_n \mathbf{v}_t] = 0 . \tag{3.4.15}$$

The set of Eqs. (3.4.10) through (3.4.15) provides us with all types of magnetohydrodynamic transition layers. Let us first consider the case where no particle transport exists across the transition layer; this is called a "discontinuity". This case is further divided into two types depending on whether the normal magnetic component exists or not.

In the case where

$$v_{1n} = v_{2n} = 0, \quad B_{1n} = B_{2n} \equiv B_n(\neq 0), \tag{3.4.16}$$

we have from Eqs. (3.4.10) through (3.4.15)

$$[\mathbf{B}_t] = 0, \quad [\mathbf{v}_t] = 0, \quad [p] = 0. \tag{3.4.17}$$

Thus, the magnetic field, the velocity and the pressure are continuous. The density, however, can have different values. Such a discontinuity is called a "contact discontinuity".

In the case where

$$v_{1n} = v_{2n} = 0, \quad B_{1n} = B_{2n} = 0, \tag{3.4.18}$$

we have from Eqs. (3.4.10) through (3.4.15)

$$[p + \frac{B_t^2}{2\mu_o}] = 0. \tag{3.4.19}$$

The tangential velocity, the tangential component of the magnetic field, the density and the pressure can all be discontinuous. Such a discontinuity is called a "tangential discontinuity".

Now we shall consider the case where there is net particle transport across the transition layer; this may be called a "shock". This case contains three types of shocks. The first type is the one in which the normal velocity, and hence the density, is continuous. Namely,

$$v_{1n} = v_{2n} \equiv v_n(\neq 0), \quad n_1 = n_2 \equiv n . \tag{3.4.20}$$

From Eqs. (3.4.11) and (3.4.15) we obtain

$$v_n = \pm \frac{B_n}{\sqrt{\mu_0 nm}} . \tag{3.4.21}$$

Using Eqs. (3.4.10) and (3.4.15), with the help of Eqs. (3.4.20) and (3.4.21), Eqs. (3.4.12) and (3.4.13) are reduced to

$$[p] = 0 \tag{3.4.22}$$

and

$$[B_t^2] = 0 . \tag{3.4.23}$$

From these results it is said that the tangential component of the magnetic field and, hence, the tangential velocity, can rotate across the layer. From Eqs. (3.4.14) and (3.4.23), however, the magnitude of the magnetic field remains unchanged. Equation (3.4.21) indicates that this transition is associated with the shear Alfveń wave, see Section 2.5. These shocks are called "Alfveń shocks" or "intermediate shocks". Since all thermodynamic quantities such as the density, temperature and pressure are continuous, they are also called "rotational discontinuities".

The other types are classified by

$$n_1 m v_{1n} = n_2 m v_{2n} \equiv F_p (\neq 0) , \quad v_{1n} \neq v_{2n} . \tag{3.4.24}$$

Hence, from Eq. (3.4.10),

$$n_1 \neq n_2 . \tag{3.4.25}$$

Combining Eqs. (3.4.10), (3.4.11), (3.4.14) and (3.4.15) yields

$$[v_n \mathbf{B}_t] = \frac{B_n^2}{\mu_0 F_p} [\mathbf{B}_t] . \tag{3.4.26}$$

This relation indicates that $\mathbf{B}_{1t} \,||\, \mathbf{B}_{2t}$, hence the magnetic field remains in the same plane normal to the transition layer which is called "coplanarity". Elimination of the $\mathbf{v}_t$ terms in Eq. (3.4.13) with the help of Eq. (3.4.11) yields

$$\frac{1}{2} F_p [v_n^2] + \frac{\gamma}{\gamma - 1} [p v_n] + \frac{1}{\mu_0} [B_t^2 v_n] - \frac{B_n^2}{2\mu_0^2 F_p} [B_t^2] = 0 . \tag{3.4.27}$$

Using the relation that $[AB] = \overline{A}[B] + \overline{B}[A]$ where $\overline{A} = (A_1 + A_2)/2$,

Eqs. (3.4.27) can be simplified with help of Eqs. (3.4.12) and (3.4.26) to

$$\frac{1}{\gamma-1}\bar{v}_n[p]+\frac{\gamma}{\gamma-1}\bar{p}[v_n]+\frac{1}{4\mu_o}[v_n][B_t]^2=0. \tag{3.4.28}$$

From Eqs. (3.4.11) and (3.4.12) we also have

$$[v_t]=\frac{B_n}{\mu_o F_p}[B_t] \tag{3.4.29}$$

and

$$F_p[v_n]+[p]+\frac{1}{2\mu_o}[B_t^2]=0. \tag{3.4.30}$$

Equation (3.4.26) can be rewritten in a scalar form as

$$[v_n B_t]=\frac{B_n^2}{\mu_o F_p}[B_t]. \tag{3.4.31}$$

Equations (3.4.28) through (3.4.31) provide a closed set of shock conditions.

Eliminating $[v_n]$ from Eqs. (3.4.28) and (3.4.30), we have

$$\left(\frac{F_p\bar{v}_n}{\gamma-1}-P\right)[p]=P\frac{[B_t^2]}{2\mu_o}, \tag{3.4.32}$$

where

$$P=\frac{\gamma}{\gamma-1}\bar{p}+\frac{1}{4\mu_o}[B_t]^2. \tag{3.4.33}$$

In the shock layer the bulk flow energy is converted into the thermal energy, so that

$$[p]>0. \tag{3.4.34}$$

Therefore, when

$$\frac{F_p\bar{v}_n}{\gamma-1}>P, \tag{3.4.35}$$

then

$$[B_t^2] > 0. \tag{3.4.36}$$

On the other hand, when

$$\frac{F_p \overline{v}_n}{\gamma - 1} < P, \tag{3.4.37}$$

then

$$[B_t^2] < 0. \tag{3.4.38}$$

The former shocks are called "fast shocks" and the latter "slow shocks". In fast shocks, the magnetic field is strengthened across the shock front, while it is weakened in the slow shocks.

Formation of the Bow Shock and the Magnetosphere

In the solar wind the interplanetary magnetic field is generally weak in the sense that the solar wind velocity v_s is much higher than the Alfveń velocity, i.e., $v_s >> B_s/\sqrt{\mu_0 n_s m_s}$. The subscript s refers to the solar wind. Therefore, Eq. (3.4.21) may not be satisfied; hence, we will ignore the Alfveń shocks. The shocks of principal concern then can be described by Eqs. (3.4.28) through (3.4.31). From Eqs. (3.4.28), (3.4.10) and (3.4.30) we obtain, assuming also $v_s >> \sqrt{\gamma p_s/n_s m_s}$ and $p_2 >> B_{2t}^2/2\mu_0$,

$$v_{2n} \approx \frac{\gamma-1}{\gamma+1} v_{sn}, \quad n_2 \approx \frac{\gamma+1}{\gamma-1} n_s, \quad p_2 \approx \frac{2}{\gamma+1} n_s m_s v_{sn}^2. \tag{3.4.39}$$

These are the conditions of strong ordinary fluid shocks. The ratio of the magnetic field can be obtained, from Eq. (3.4.31) by setting the right hand side to zero, as

$$B_{2t} \approx (v_s/v_{2n}) B_{st} \approx \frac{\gamma+1}{\gamma-1} B_{st}, \tag{3.4.40}$$

where B_{st} is the transverse component of the solar wind magnetic field. The magnetic field is thus increased in the post-shock flow, indicating that the shock of this type possesses the nature of a fast shock. This shock is called the "bow shock".

Equation (3.4.29) gives the increase in the transverse velocity:

$$v_{2t} - v_{st} \approx \frac{2}{\gamma-1} \frac{B_{sn} B_{st}}{\mu_0 n_s m_s v_s}. \tag{3.4.41}$$

Relations (3.4.39), (3.4.40), and (3.4.41) are not sufficient to determine the position and shape of the shock front unless the downstream boundary conditions are specified.

What we know is that in the absence of any interaction with the solar wind plasma the geomagnetic field would form an earth-centered dipole structure. As we have described in the first part of this section, the geomagnetic field is so strong near the earth's surface that the solar wind must be completely blocked somewhere far from the earth's surface. At the position where the solar wind is stopped by the geomagnetic field, the condition (3.4.18) may well be satisfied. In terms of the discontinuities discussed in the foregoing subsection, the transition layer (magnetopause) between the solar wind plasma and the territory of the geomagnetic field can be a tangential discontinuity. Thus, the only relation that can determine the position and shape of the magnetopause is Eq. (3.4.19).

The upstream conditions of the magnetopause are continuously connected to the downstream conditions of the bow shock we have just discussed. Furthermore, the geomagnetic field on the downstream side of the magnetopause is also largely deformed by this interaction. Therefore, the shapes and positions of the bow shock and the magnetopause are difficult to determine self-consistently by an analytical method, unless an oversimplified assumption such as two-dimensionality is made. The most viable method at hand may be a magnetohydrodynamic (MHD) simulation. Numerical solutions will be given later, but, in advance, we shall make a rough estimate of the distance between the shock and the magnetopause (stagnation point of the solar wind). This distance is called the "stand-off distance."

Let us assume that the magnetopause and the shock take concentric spherical surfaces near the sun-earth axis. Spherical coordinates are then chosen in such a way that the origin is located at the center of the earth and the axis points to the sun. On the shock the solar wind flow suddenly deflects. Thus, a vorticity Ω is generated behind it as is shown in Fig. 3.4. Assuming an axisymmetric flow, we can introduce a stream function Ψ of a flow generated by the vorticity on the shock. The radial and polar-angle components of the flow are described as

$$\left.\begin{aligned} v_R &= -\frac{1}{R^2\sin\Theta}\frac{\partial\Psi}{\partial\Theta} \\ v_\Theta &= \frac{1}{R\sin\Theta}\frac{\partial\Psi}{\partial R}. \end{aligned}\right\} \qquad (3.4.42)$$

The vorticity equation is thus given by

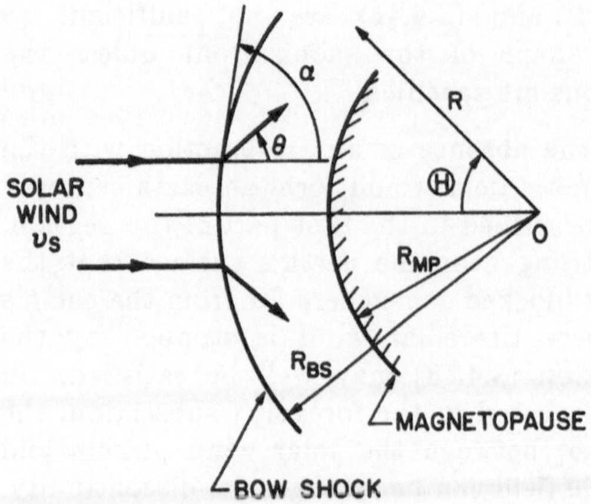

Fig. 3.4. Geometry showing the relationship of the solar wind direction, the bow shock position and the magnetopause position

$$\frac{1}{R\sin\Theta}\frac{\partial^2\Psi}{\partial R^2}+\frac{1}{R^3}\frac{\partial}{\partial\Theta}\left(\frac{1}{\sin\Theta}\frac{\partial\Psi}{\partial\Theta}\right)=\Omega. \tag{3.4.43}$$

Since the vorticity in an axisymmetric constant-density flow is proportional to $R\sin\Theta$, Ω in Eq. (3.4.43) may be given by

$$\Omega=\frac{v_s}{R_c^2}R\sin\Theta, \tag{3.4.44}$$

where v_s is the solar wind speed and R_c is the curvature of the solar wind flow deflected on the shock. The curvature R_c of the flow on the shock can be estimated as follows: Denoting the jump of the normal flow speed across the shock by ϵ, we have $v_{2n}=\epsilon v_{1n}$ and $v_{2t}=v_{1t}$. With help of Fig. 3.4, we obtain

$$\tan\alpha=\epsilon\tan(\alpha-\theta),$$

where α is the angle of the shock tangent with respect to the upstream flow direction and θ is the angle of the flow deflection on the shock. From this relation we obtain

$$\frac{d\theta}{d\alpha}=1-\frac{1}{\epsilon}\frac{\cos^2(\alpha-\theta)}{\cos^2\alpha}.$$

Since we are concerned with a near-axis flow, we can put $\theta\approx 0$, hence, $d\theta/d\alpha\approx(\epsilon-1)/\epsilon$. The curvature R_c, therefore, may be approximated by

$$R_c^2 \approx R_{BS}^2 \epsilon^2/(1-\epsilon)^2 ,$$

where R_{BS} is the radial distance of the bow shock. Using this relation in Eq. (3.4.44) we obtain

$$\Omega = \frac{(1-\epsilon)^2 v_S}{\epsilon^2 R_{BS}^2} R \sin\Theta . \tag{3.4.45}$$

Thus, assuming a solution of the form $\Psi = \sin^2\Theta \sum_{n=-\infty}^{\infty} a_n R^n$ and using a boundary condition of $v_{2t} = v_{1t}$ at $R = R_{BS}$, Eq. (3.4.43) is solved as

$$\Psi = \frac{v_S R_{BS}^2 \sin^2\Theta}{30\epsilon^2} \left[3(1-\epsilon)^2 \left(\frac{R}{R_{BS}}\right)^4 - 5(1-4\epsilon)\left(\frac{R}{R_{BS}}\right)^2 + 2(1-\epsilon)(1-6\epsilon)\left(\frac{R_{BS}}{R}\right) \right] . \tag{3.4.46}$$

The condition that $\Psi = 0$ on the magnetopause ($R = R_{MP}$) yields the stand-off distance, $\Delta = R_{BS} - R_{MP}$, as

$$\Delta \approx \epsilon R_{BS} (1 - \sqrt{8\epsilon/3} + 3\epsilon) , \tag{3.4.47}$$

where $\epsilon = v_{2n}/v_{1n} = (\gamma-1)/(\gamma+1)$ for a strong shock, see Eq. (3.4.39). For example, we have for $\gamma = 5/3$ and $R_{MP} = 10 R_E$

$$\Delta \approx 3 R_E . \tag{3.4.48}$$

As we have seen above, rough estimates of the magnetopause position (stagnation point) and the shock position on the solar side can be obtained analytically. However, the entire structure and the quantitative features can not be elucidated in an analytical fashion. In the following we shall show some stationary solutions of the solar wind-magnetosphere interaction obtained by means of a high-precision, three-dimensional MHD simulation code.

Figure 3.5 illustrates an example of stationary state solar wind streamlines and the geomagnetic field lines in the noon-midnight meridian plane resulting from the interaction between the solar wind and the magnetosphere. In this example the solar wind speed was 300 km/sec and the density was $5 \times 10^6 \mathrm{m}^{-3}$. The magnetic pressure

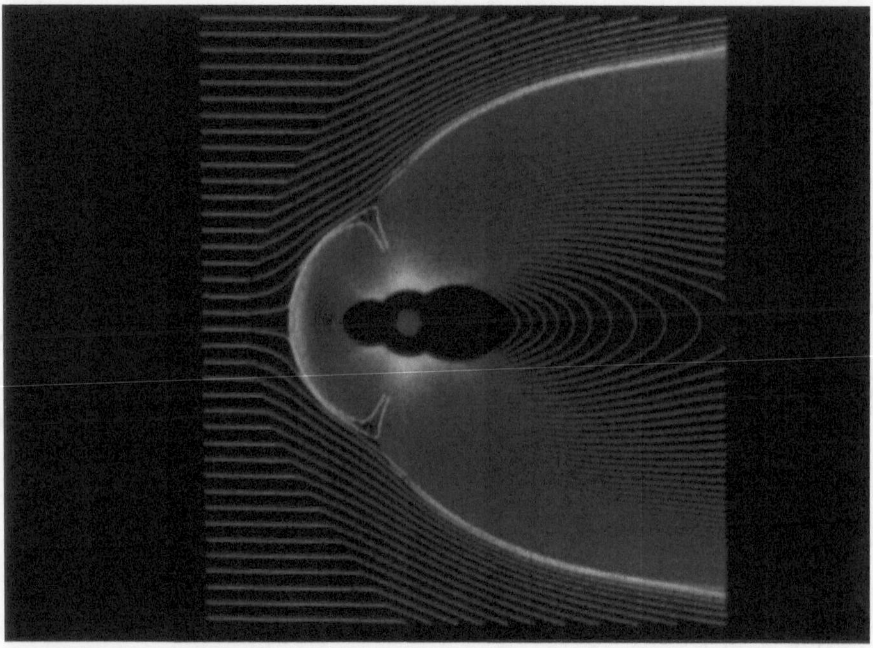

Fig. 3.5. Quasi-stationary solar wind streamlines and the geomagnetic field lines in the noon-midnight meridian plane: A result of 3D MHD simulation with a simulation box of $20R_E \geq X \geq -30R_E$, $25R_E \geq Y \geq -25R_E$ and $30R_E \geq Z \geq -30R_E$

in the magnetosphere was initially 3.2×10^{-11} Newton/m^2. The magnetic field of the solar wind was ignored. Illustrated in Fig. 3.6 are streamlines and field lines mapped on the equatorial plane of the same example. Three typical features are observable in these figures. First, the uniformly blowing solar wind suddenly deflects at a parabolic plane on the solar side of the earth, which turns out to be a fast shock (bow shock). Second, the geomagnetic field lines are compressed on the solar side and blown towards the tail forming the magnetosphere. The third feature is the formation of the northern and southern cusp-like dividing regions of the magnetic field lines at dayside high latitudes.

A three-dimensional shape of the dayside magnetosphere is shown in Fig. 3.7. The cross-section of the magnetosphere is rather prolate in the north-south direction, say, $50R_E$ in north-south direction and $40R_E$ in dawn-dusk direction. The magnetopause distance on the dayside is about $10R_E$.

In the simulation an adiabatic law, $pn^{-\gamma} = \text{const.}$, is employed. In the example shown in Fig. 3.5 and Fig. 3.6, $\gamma = 5/3$ was used. In order

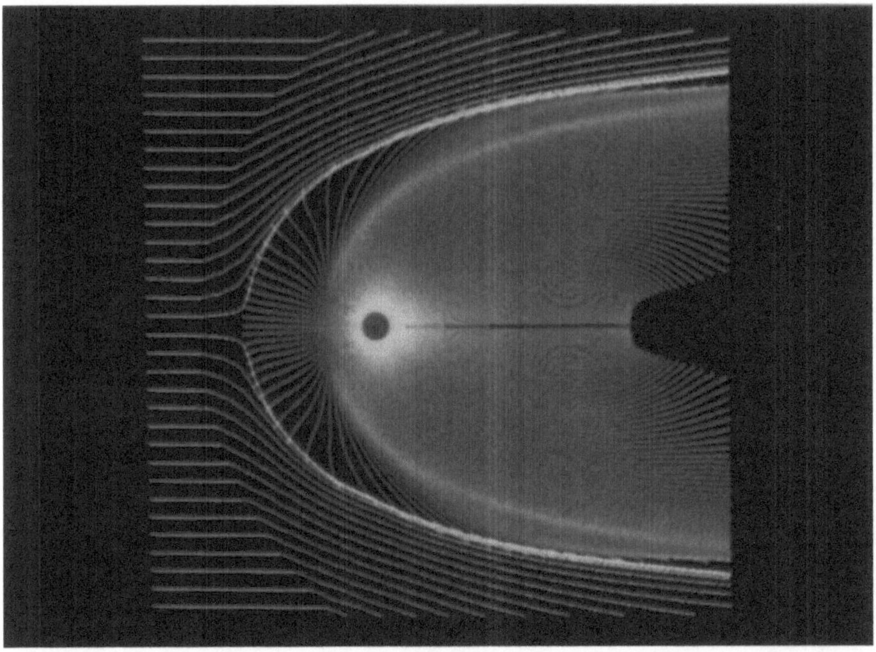

Fig. 3.6. Quasi-stationary solar wind streamlines on the equatorial plane and the geomagnetic field lines mapped on the equatorial plane for the same conditions as Fig. 3.5

to examine the dependence of the stand-off distance (distance between the shock and the magnetopause) on the adiabatic constant (γ), simulation runs were repeated for different values of γ keeping the other conditions the same. Figure 3.8 gives the stand-off distance as a function of γ. The dotted line represents the theoretical curve previously derived for a strong shock, i.e., Eq. (3.4.47). Comparing this with the observed stand-off distance, we are able to estimate the value of γ of the solar wind.

For a more quantitative profile of the bow shock and magnetosphere, we plot in Fig. 3.9 the distributions of the plasma pressure, the temperature, the density, the flow velocity, and the magnetic field along the sun-earth line. As is clear in this figure, at the shock the pressure and the density suddenly jump to high values, while the solar wind speed suddenly drops. From these features we can conclude that the shock is indeed a fast shock, see, Eqs. (3.4.34)-(3.4.36). Coming closer to the earth, we encounter another discontinuity at about $r = 10R_E$. At this discontinuity the pressure and the density drop sharply to almost nothing, while the magnetic field rises abruptly. More importantly, the solar wind speed

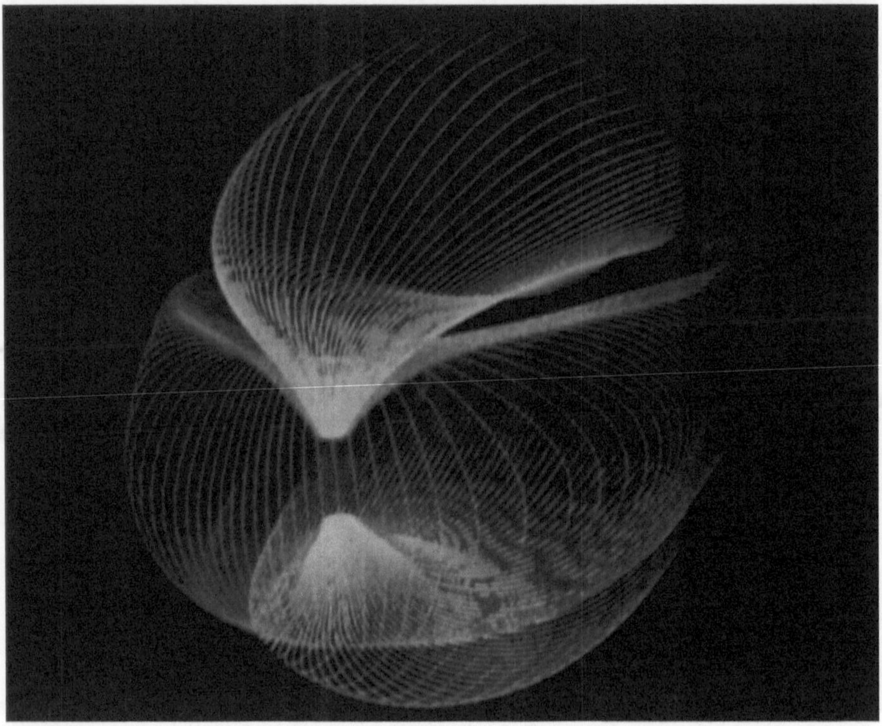

Fig. 3.7. Three-dimensional geomagnetic field lines originating from the geomagnetic latitudes of 68 degree (green) and 72 degree (yellow) for the same conditions as Fig. 3.5

completely vanishes at the discontinuity. These features are the very features of a tangential discontinuity, i.e., Eqs. (3.4.18) and (3.4.19). This discontinuity corresponds to the magnetopause. Another interesting and significant feature of Fig. 3.9 is an indication of the formation of a high pressure region in the tail region. The pressure sharply increases at $r \approx 8R_E$ on the night side and keeps its value down in the tail.

An additional feature observed in Figs. 3.5 and 3.6 is that the solar wind can never get inside the magnetosphere. This suggests that some special (kinetic) transport processes must be invoked for a continuous entry of the solar wind energy into the magnetosphere.

Magnetotail Formation

The distant geomagnetic field lines are squeezed towards the downstream side by the solar wind. It is expected therefore that the magnetopause will extend far in the downstream direction like a

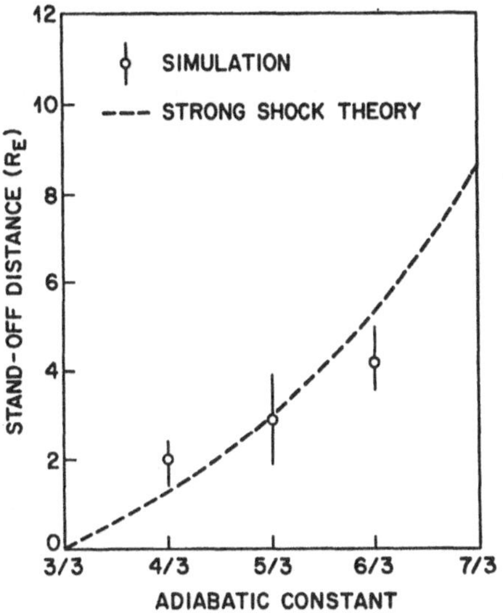

Fig. 3.8. The stand-off distance of the bow shock as a function of the adiabatic constant obtained by the MHD simulation

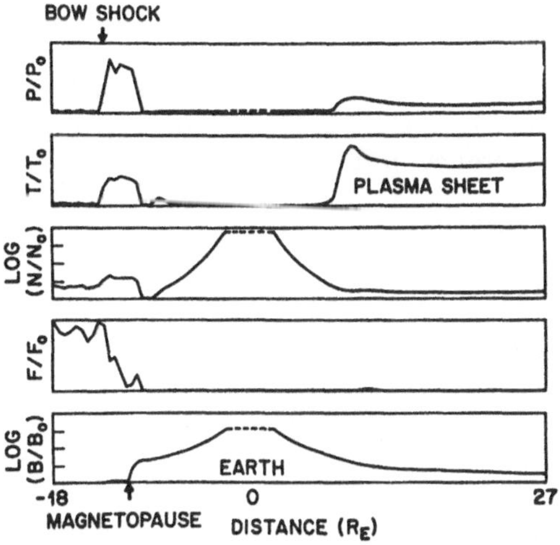

Fig. 3.9. Distributions of the plasma pressure, the temperature, the density, the solar wind momentum and the northward magnetic field on the sun-earth line (same simulation conditions as Fig. 3.5)

comet. The night side magnetosphere which extends far into the solar wind is called the "magnetospheric tail", "magnetotail", or simply "tail". Since all the distant geomagnetic field lines are confined in a limited volume of the tail, which otherwise would have expanded to an infinite volume, the energy density of the geomagnetic field can be enormously enhanced there. Suppose that the magnetosphere is empty of plasma; hence, no currents can flow inside. Then an electric current must flow only on the magnetopause to shield the magnetic field from the solar wind. This current will be the diamagnetic type of current produced by the pressure of the solar wind plasma. Such a surface current is called a "Chapman-Ferraro current".

When the magnetopause is an ideal tangential discontinuity, no particle flux can exist across it (see Figs. 3.5 and 3.6). Any magnetospheric plasma must then be solely of earth origin. The plasma pressure of earth origin, that is, the ratio of the particle pressure of earth origin to the geomagnetic pressure, is extremely low everywhere in the magnetosphere. Thus, the plasma current in the magnetosphere, namely, the diamagnetic current of earth origin, is negligible, so that the magnetic field structure in the magnetosphere is determined by the sum of the dipole field and the field produced by the Chapman-Ferraro current. Given the magnetopause shape, the structure is determined by solving the Laplace equation for the magnetic potential Ω_m with the boundary condition $\Omega_m = 0$ on the magnetopause;

$$\nabla^2 \Omega_m = 0, \qquad (3.4.49)$$

where $\Omega_m = \Omega_d + \Omega_{CF}$, Ω_d being the earth's dipole field potential and Ω_{CF} the potential due to Chapman-Ferraro current. One calculated field line structure is shown, for reference, in Fig. 3.10.

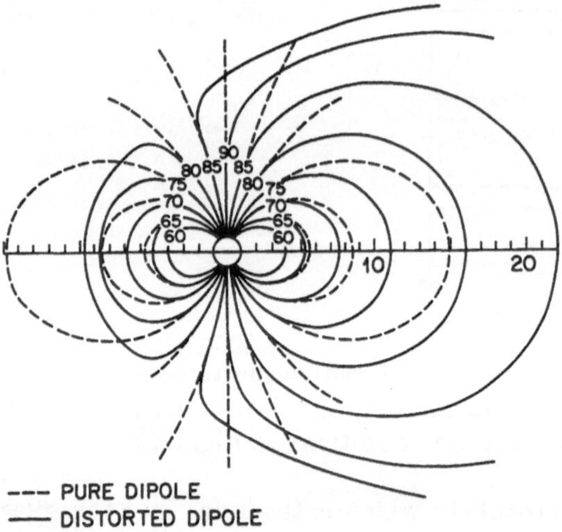

Fig. 3.10. Dipole field lines (*dotted*) and the deformed field lines (*solid*) by the Chapman-Ferraro current (G. C. Mead, *J. Geophys.* *69*, 1181, 1964).

The observed magnetic field structure is very different from this, particularly in the distant tail. The observed magnetic field lines in the tail beyond $r \sim 20R_E$ become nearly parallel to the solar wind flow. Such a configuration cannot be produced by the Chapman-Ferraro current alone. To account for the observed structure, a plasma current must flow somewhere inside the magnetosphere. In other words, the internal plasma pressure must be as large as the magnetic pressure there. Since the geomagnetic field lines reverse across the equatorial plane, the current as a whole is expected to flow from the morning (dawn) side to the evening (dusk) side in the equatorial region. The region of high plasma pressure in the equatorial tail region is called the "plasma sheet", and the dawn-dusk current is called the "neutral-sheet current" or "tail current".

This fact reminds us of a very crucial problem in plasma physics: How can such a hot and/or dense plasma be created in the midst of the magnetotail? Suppose that the magnetic field strength is 30 nT (nano tesla). Then the magnetic pressure is about 3.6×10^{-10} newton/m^2. In order to balance this magnetic pressure, the plasma density n times the temperature T must be 10^{13}K m^{-3}.

Let us first assume that the plasma to supply this high pressure plasma is solely of earth origin. Most of the field lines in the tail region are connected to the terrestrial polar regions. Plasma particles in the tail region are therefore considered to be those escaping along field lines from the polar atmosphere (polar wind). The temperature and the density may be estimated by a treatment similar to that in Section 3.3. This estimates the density and temperature to always be $n \lesssim 10^5$ m^{-3} and $T \lesssim 10^4$K. These values are several orders of magnitude smaller than the pressure required to produce an antiparallel field structure in the equatorial tail. As we have seen in Fig. 3.10, the Chapman-Ferraro current itself cannot be so strong as to largely compress the dipole field near the equatorial plane. Therefore, if the tail plasma is purely of earth origin, an exceptional mechanism must exist that could sweep all the plasma particles in the entire tail region towards the equatorial region and heat them to an extremely high temperature.

Another possibility is to invoke entry of the solar wind particles into the equatorial region of the tail. Since the energy density of the solar wind plasma is of the same order as the required energy density, this possibility is much more likely, at least energetically, than the above concept of an earth origin. However, as we have already learned, binary collisions among particles are too infrequent to scatter sufficient solar wind particles into the magnetosphere. Indeed, the classical transport time scale is about a year. Therefore, an anomalously swift particle transport process across the magnetopause must be invoked, which is a challenging problem in plasma physics. Another problem is to find a reasonable process that will act to sweep particles entering the magnetosphere towards the equatorial plane of the tail. Possibilities will be discussed in Volume II of this series. But, for the time being, we assume the existence of sufficient hot particles in the equatorial tail region and examine the stationary tail structure.

In the distant tail, say, $r > 20R_E$, it is likely that the field lines are almost parallel to the solar wind. The direction of the field line reverses across the geomagnetic equatorial plane where the magnetic field is almost null. Thus, we call this magnetically null plane the "neutral sheet". From the standpoint of plasma physics, the neutral sheet region of the magnetotail is the most interesting place in the magnetosphere, because the plasma energy density is very high (high β plasma in terms of plasma physics) and quantities such as the plasma temperature, density and the magnetic field change drastically in a direction perpendicular to the neutral sheet. Thus, this region is expected to be subject to plasma instabilities both macroscopically and microscopically. The microinstability is expected to induce anomalous dissipation (see Vol. II) and an ideal MHD structure may not be steadily sustained. To study these behaviors theoretically, it is important first to construct an analytic equilibrium solution.

Let us take a rectangular coordinate system (x, y, z) where the x-y plane represents the neutral sheet, the x axis points to the sun, the y axis towards dusk (west) and, hence, the z axis to the north. The equilibrium magnetic field **B** has only an x component which is variable only in the z direction, i.e., $\mathbf{B} = [B_x(z), 0, 0]$. The corresponding vector potential **A** has only a y component, A_y, and is related to **B** through $B_x = -dA_y/dz$.

In this configuration of the magnetic field we have two constants of motion, the Hamiltonian H, and the y component of the canonical momentum P_y.

From Eqs. (1.2.13) and (1.2.11) we have

$$H = \frac{1}{2} m_j v_j^2 + e_j \phi \tag{3.4.50}$$

$$P_y = m_j v_{yj} + e_j A_y , \tag{3.4.51}$$

where ϕ is an electric potential, and the subscript j refers to either electrons (e) or protons (p). Using these constants of motion, the distribution function f_j for species j can be expressed by

$$f_j = N_o\left(\frac{m_j}{2\pi T_j}\right)^{3/2} \exp\left(-\frac{H}{T_j} + \frac{u_j}{T_j}P_y - \frac{m_j}{2T_j}u_j^2\right), \tag{3.4.52}$$

where N_o and u_j are constant values. Integrating this equation over velocity space, we obtain

$$n_j = N_o \exp\left[-(e_j\phi - e_j u_j A_y)/T_j\right].$$

The charge neutrality condition yields

$$\phi = 0, \ \frac{eu_p}{T_p} = -\frac{eu_e}{T_e} = C.$$

Thus, we have

$$n(z) = N_o e^{CA_y(z)}. \tag{3.4.53}$$

From $\nabla \times \mathbf{B} = \mu_o \mathbf{J}$, we have

$$\frac{dB_x}{dz} = \mu_o \sum_j e_j \int v_{yj} f_j dv_j. \tag{3.4.54}$$

Differentiation of Eq. (3.4.54) with respect to z yields

$$\frac{d^2B_x}{dz^2} = -CB_x\frac{dB_x}{dz}.$$

With the boundary conditions that

$$B_x(0) = 0, \ B_x(\pm\infty) = \pm B_o,$$

this differential equation is solved to give

$$B_x(z) = B_o \tanh\left(\frac{z}{L}\right), \tag{3.4.55}$$

where the constant C is replaced with $2/B_oL$. Using this solution to Eq. (3.4.53) we obtain

$$n(z) = N_o \operatorname{sech}^2\left(\frac{z}{L}\right). \tag{3.4.56}$$

The neutral sheet current and the pressure are thus given by

$$J_y(z) = \frac{B_o}{\mu_o L}\,\mathrm{sech}^2\!\left(\frac{z}{L}\right) \qquad (3.4.57)$$

$$p(z) = \frac{B_o^2}{2\mu_o}\,\mathrm{sech}^2\!\left(\frac{z}{L}\right). \qquad (3.4.58)$$

These solutions, i.e., Eqs. (3.4.55) through (3.4.58), are called the Harris equilibrium, and provide us with one analytical equilibrium representing the tail structure near the neutral sheet. We note that u_j in Eq. (3.4.52) corresponds to the diamagnetic drift.

The fact that the axis of the earth's magnetic moment is perpendicular to the direction of the solar wind makes it difficult to analytically obtain a three dimensional magnetosphere equilibrium. Because of the real three dimensionality, the Grad-Shafranov type of analysis described in Chapter 1 is not appropriate. The most viable way of obtaining the magnetotail structure may be an MHD simulation such as that which was used in obtaining the bow shock and the dayside magnetosphere (see Figs. 3.5 and 3.6).

In an MHD simulation, however, we cannot evaluate the particle transport of the solar wind plasma through the magnetopause boundary. As we have seen in Figs. 3.5 and 3.6, neither solar wind particles nor momentum and energy can enter into the magnetosphere in an ideal condition. Thus, in the simulation to obtain the magnetotail formation we must use a larger simulation box and repeat parameter runs for cases where the plasma content in the magnetosphere has different values. Figure 3.11 illustrates a three dimensional magnetic structure of the magnetosphere when the initial plasma pressure beyond $r = 10R_E$ is assigned to be 3.2×10^{-11} newton/m^2 and the solar wind speed (unmagnetized) is 300 km/sec. The tail pressure at the equator reaches about 3×10^{-10} newton/m^2 which is comparable to the dynamic pressure of the solar wind ($\sim 3.6 \times 10^{-10}$ newton/m^2). Figure 3.12 is a composite display of the meridian and equatorial cross sections of the magnetosphere for the same run as Fig. 3.11. Figure 3.13 depicts the three dimensional magnetic structure when the initial plasma pressure beyond $r = 10R_E$ is reduced to half of that in Fig. 3.12. Comparison of Figs. 3.11 and 3.13 tells us that the dimension of the tail cross-section and the width of the plasma sheet become smaller as the initial magnetospheric plasma pressure is made smaller. This indicates that by carefully comparing the simulation parameter with the observational data we can estimate how many solar wind particles must be deposited in the magnetosphere.

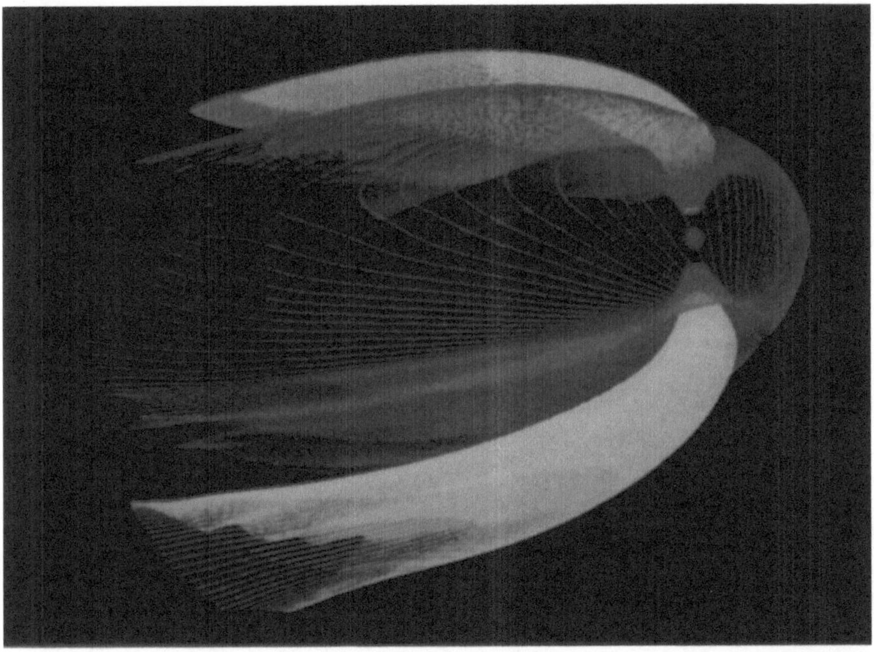

Fig. 3.11. Same as Fig. 3.7 except for a simulation box of 20 $R_E > X > -60R_E$, $30\ R_E \geq Y \geq -30R_E$ and $35\ R_E \geq Z \geq -35R_E$

Convection of the Magnetospheric Plasma

So far we have ignored the plasma flow in the magnetosphere. Viewed from a rest frame in the solar system, the earth is rotating. The magnetospheric plasma of earth origin will have an angular velocity of corotation in this frame. This means that particles in the inner magnetosphere execute circular convection relative to the magnetopause. Furthermore, as we have seen in the discussion of magnetotail formation, it is likely that the magnetopause cannot be an ideal tangential discontinuity, but that the shielding it provides is broken sporadically or continuously on at least a part of the surface. Let us now assume that there is a steady entry of the solar wind momentum into the magnetosphere and study a stationary structure of the magnetosphere resulting from it.

From $\mathbf{E} = -\mathbf{v} \times \mathbf{B}$, i.e., Eq. (1.6.15), the existence of convection is equivalent to the existence of an electric field in the rest frame. When it is an ideal tangential discontinuity, the magnetopause is a constant electric potential surface; therefore, no electric field will appear inside the magnetosphere. Conversely speaking, the existence of a steady momentum entry implies that the electric potential on the

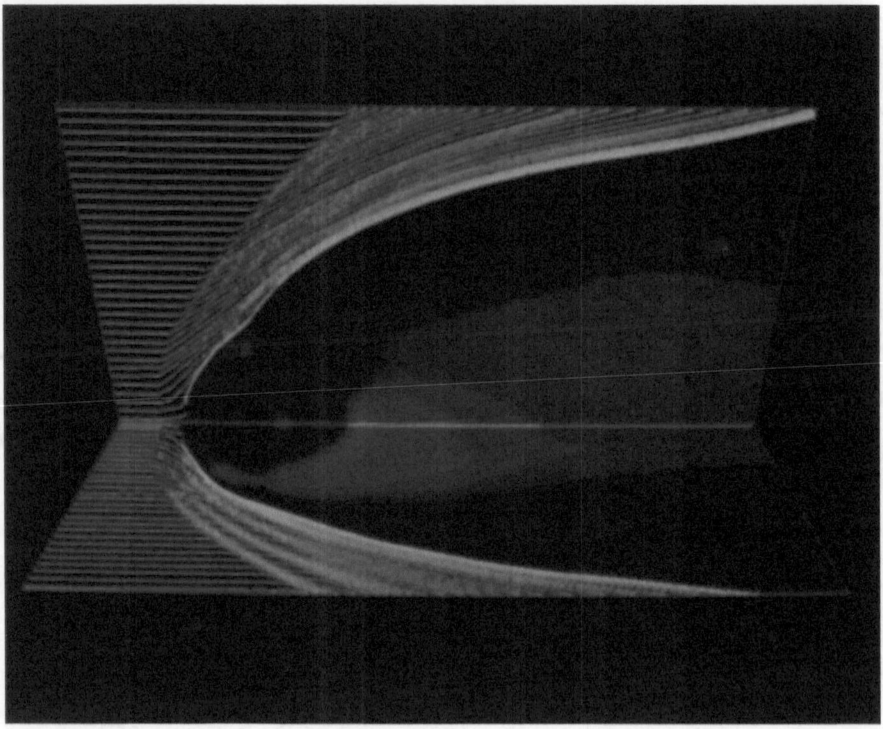

Fig. 3.12. A composite display of the solar wind stream lines, the magnetosphere boundaries, and the plasma temperature distributions in the meridian and equatorial planes for the same condition as Fig. 3.11

magnetopause is no longer uniform. This indicates that an electric field will appear in the magnetosphere in association with momentum transport through the magnetopause. Since it is the solar wind that produces the electric field inside the magnetosphere whatever the transport mechanism is, the electric field direction in the magnetosphere must be dawn to dusk in a stationary state.

This can be easily understood when we look at the equatorial plane. The transported momentum just inside the magnetopause must have the same direction as that of the solar wind, resulting in a dynamo electric field which points from the magnetosphere to the solar wind on the dawn side (where the geomagnetic field is north) and opposite on the dusk side. The transported momentum must be decreasing and disappear as we go inside. Thus, from $\nabla \cdot \mathbf{E} = -\rho/\epsilon_0$, positive charges will accumulate just inside the dawn side magnetopause and negative charges just inside the dusk side magnetopause. These

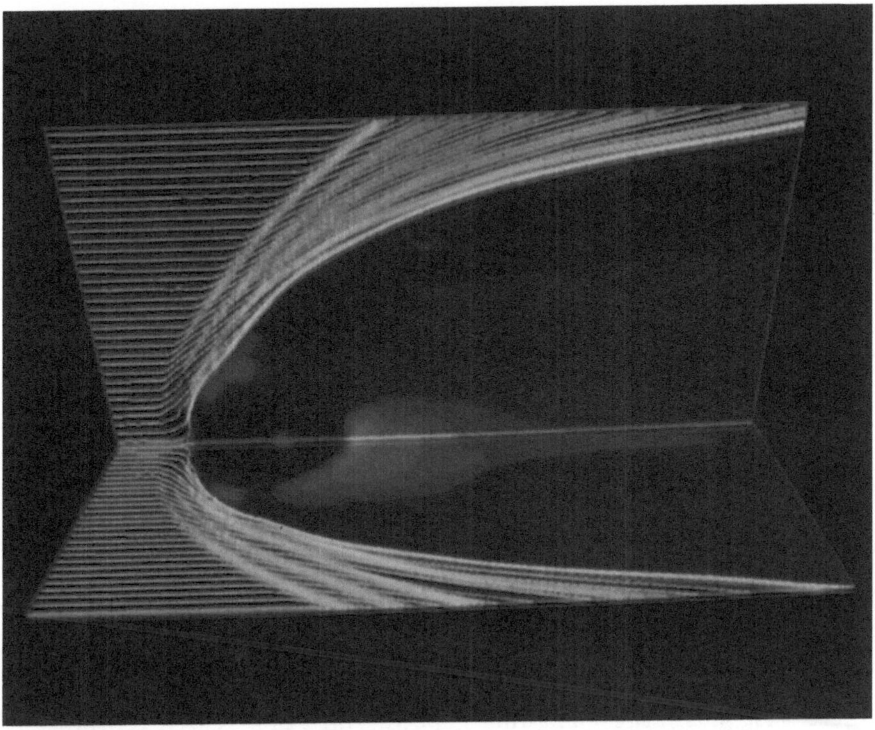

Fig. 3.13. Same as Fig. 3.11 but when the initial plasma pressure in the magnetosphere is halved

charges will produce a dawn-dusk electric field across the equatorial magnetosphere except for the (thin) transport layer just inside the magnetopause. The dawn-dusk electric field thus produced will cause magnetospheric plasma from the tail to drift towards the dayside magnetopause. The plasma will turn at the maximum or minimum position of the potential and flow out towards the tail through the transport layer. In reality, the transport layer will be a complicated structure where plasma waves are strongly excited or interplanetary magnetic field lines are connected to the geomagnetic field lines by magnetic reconnection. Therefore, the magnetospheric particles drifting to the transport region are substantially lost into the solar wind.

In a steady state, therefore, a magnetospheric plasma particle is subject to a drift motion under the action of an electric field and the geomagnetic field. The major electric fields consist of the earth's corotation field ($\mathbf{E}_{\text{rot}}$) and the dawn-dusk electric field ($\mathbf{E}_{\text{SW}}$) which results from the incomplete shielding of the magnetopause from the solar wind.

For simplicity we shall approximate the geomagnetic field by a dipole field, Eq. (3.4.1). The ratio between the ∇B-drift and the curvature drift given by Eqs. (1.7.17) and (1.7.13) is roughly given by the ratio of perpendicular to parallel energy since the characteristic lengths are of the same order. Assuming that 90 degree pitch angle particles are more densely populated, we can disregard curvature drift and retain only the ∇B-drift as the magnetic drift. Then from Eqs. (1.7.17b) and (3.4.1) the drift motion in the magnetosphere is described by

$$\mathbf{v}_D = \frac{1}{B^2}\left[\mathbf{E}_{SW} + \mathbf{E}_{rot} - \nabla\left(\frac{\mu B}{q}\right)\right] \times \mathbf{B}, \tag{3.4.59}$$

where q is the charge and μ is the magnetic moment.

In the equatorial plane the corotation electric field, $(\Omega \times \mathbf{r}) \times \mathbf{B}$, is represented by

$$\mathbf{E}_{rot} = \nabla\left(\frac{\Omega_E M}{r}\right), \tag{3.4.60}$$

where Ω_E is the angular speed of the earth's rotation, $\Omega_E = 2\pi/24 \times 3600\ \text{sec}^{-1}$, and M is the earth's dipole moment, $M = 8 \times 10^{15}$ tesla m^3. The dawn-dusk electric field is represented by

$$\mathbf{E}_{SW} = \nabla(E_0 r \sin\phi), \tag{3.4.61}$$

where E_0 is the amplitude of the dawn-dusk electric field and ϕ is the azimuthal angle from the solar direction. Substituting Eqs. (3.4.60) and (3.4.61) into Eq. (3.4.59), we have

$$\mathbf{v}_D = \frac{1}{B^2}\mathbf{B} \times \nabla V, \tag{3.4.62}$$

where

$$V = -E_0 r \sin\phi - \frac{\Omega_E M}{r} + \frac{\mu M}{q r^3}. \tag{3.4.63}$$

Equation (3.4.62) indicates that lines of constant V are equivalent to streamlines in the equatorial plane. In the distant tail the corotation and ∇B terms become negligible and the drift is governed by the dawn-dusk electric drift caused by the solar wind. The particle orbits are more or less uniform and directed towards the earth from the tail. Nearer the earth, on the other hand, the corotation and the ∇B drift become more important.

Let us first consider the drift orbits for low energy particles ($\mu = 0$). In this case a stagnation point between sunward convection and corotation appears at $\phi = 90°$ (dusk side) and at

$$r = \left(\frac{\Omega_E M}{E_0} \right)^{1/2} . \tag{3.4.64}$$

For $E_0 = 4 \times 10^{-4}$ V/m, for example, we have $r \approx 6\,R_E$. This indicates that the streamlines are separated into two groups at the separatrix passing the stagnation point. One group is the closed streamlines around the earth (corotation zone) and the other is the open streamlines representing the sunward convection. Figure 3.14(a) shows an example of streamlines for cold particles ($\mu = 0$). As can be seen from Eq. (3.4.63), the streamlines are the same for protons and electrons when $\mu = 0$.

The photoionized ionospheric plasma particles consist of low energy particles, no greater than 1 eV, so that the ∇B drift is negligibly small compared with the corotation speed, i.e.,

$$\frac{3\,\mu M}{|q|\,r^4} << \frac{\Omega_E M}{r^2}$$

for $r > R_E$. From this the photoionized ionospheric particles within a volume covered by the geomagnetic field lines threading the closed streamline zone in Fig. 3.14(a) are considered to be confined there indefinitely. The earth's ionized particles born outside of this volume (high latitudes) can, however, drift away along the open streamlines. It is expected therefore that the plasma population of earth origin can be markedly reduced beyond the corotation boundary. The corotation region is called the "plasmasphere" and the boundary is called the "plasmapause".

Let us next consider the case for hot particles where the ∇B drift cannot be neglected. In contrast to the previous case which represents the orbits of plasma particles of earth origin, this case involves plasma particles of magnetospheric or of solar origin.

For electrons ($q<0$) the direction of ∇B drift is the same as that of the corotation, so that the pattern of streamlines is essentially the same as that of Fig. 3.14(a). The streamlines for electrons with $E_0 = 4 \times 10^{-4}$ V/m and $\mu = 0$, 10, 50 eV/nT are shown in Fig. 3.15.

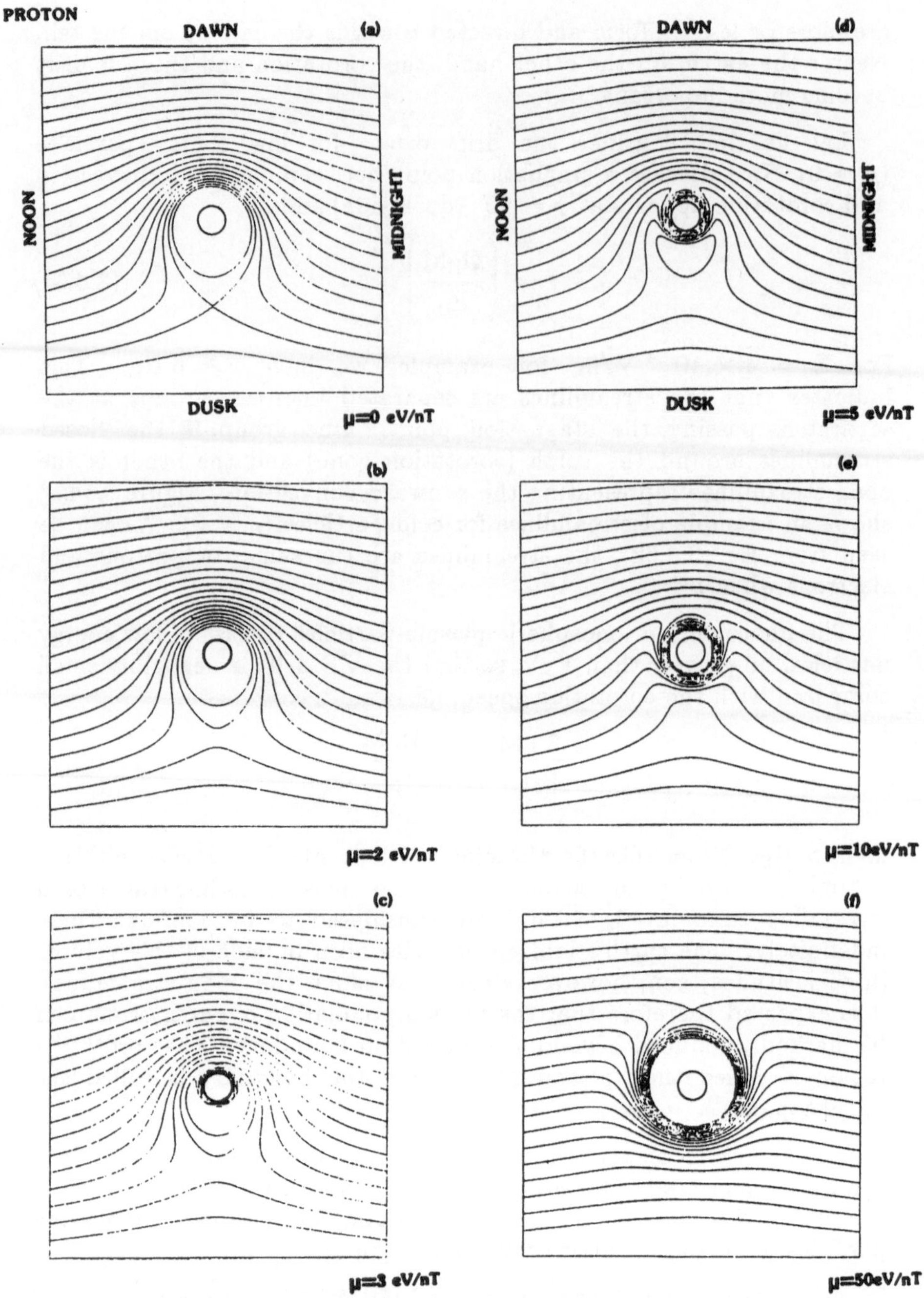

Fig. 3.14. Drift orbits of protons in the equatorial plane for different magnetic moments

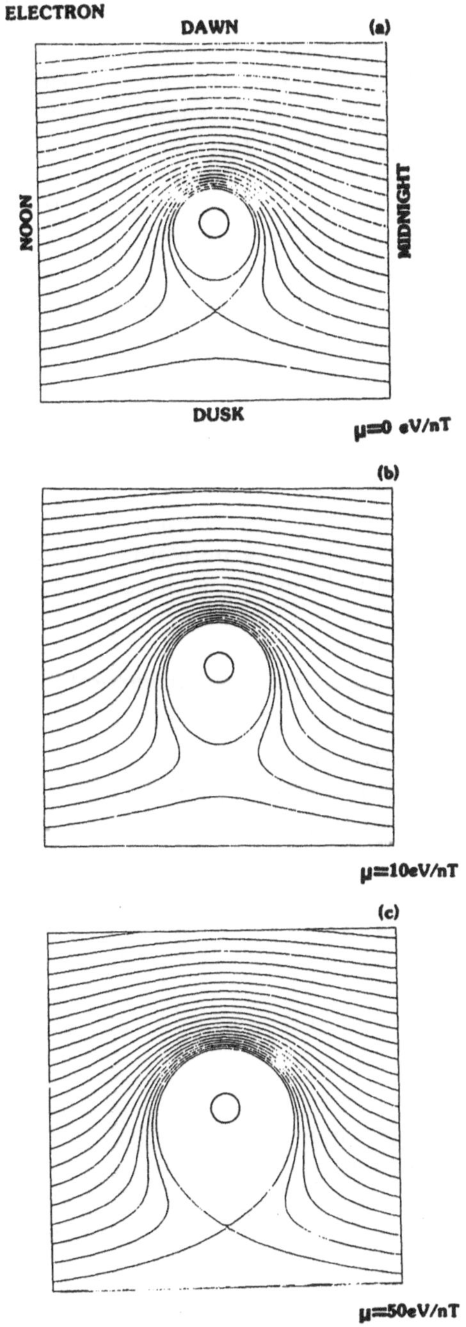

Fig. 3.15. Drift orbits of electrons in the equatorial plane for different magnetic moments

For protons (q>0), on the other hand, the direction of ∇B drift is opposite to that of corotation. Streamlines for protons with $E_0 = 4 \times 10^{-4}$ V/m and $\mu = 0, 2, 3, 5, 10, 50$ eV/nT are shown in Fig. 3.14. As can be seen from this figure, the streamline pattern begins to be modified, particularly on the dawn side, when μ becomes greater than 3 eV/nT. Protons with energy of 3 eV in the distant tail where the northward magnetic field is of the order of 1 γ(nT$=10^{-9}$ tesla) can approach as close as 2 R_E on the dawn side where the energy becomes as large as 10 keV, see Fig. 3.14(c). Protons with $\mu = 5 \sim 10$ eV/nT can circle the earth at radial distances of $r = 2 \sim 3\ R_E$, see Fig. 3.14(d) and (e). As the magnetic moment becomes as large as $\mu = 50$ eV/nT, however, a forbidden zone which the magnetospheric protons cannot access expands as large as $r = 4 \sim 6\ R_E$ as shown in Fig. 3.14(f). The stagnation point in this case turns on the dawn side.

Figure 3.16 shows an overlay of the proton and electron orbits of certain energies. As can be seen from this figure, a proton-rich region appears on the dusk side and an electron-rich region on the dawn side. This suggests that a pair of field-aligned currents (towards the ionosphere on the dusk side and away from the ionosphere on the dawn-side) can be generated. The observed field-aligned current pair called "region 2 current" may be attributed to this process.

In the above discussion we assume that shielding of the solar wind at the magnetopause is violated by some means and that a dawn-dusk electric field appears inside the magnetosphere. Whatever the mechanism may be, the entry is through the magnetopause. Thus, the magnetospheric plasma adjacent to the magnetopause must move in the same direction as the solar wind, thereby positive charges appearing on the dawn side and negative charges on the dusk side. This is because

$$\nabla \cdot \mathbf{E} = -\nabla \cdot (\mathbf{v} \times \mathbf{B}) \cong \mathbf{B} \cdot \Omega, \qquad (3.4.65)$$

where $\Omega = \nabla \times \mathbf{v}$ (vorticity) and $\mathbf{B}$ is assumed to be more or less uniform in the region concerned. If such charges arise in the magnetosphere, another pair of field-aligned currents will flow downward on the dawn and upward on the dusk side. The field-aligned current pair called "region 1 current" may be attributed to this process.

Shown in Fig. 3.17 is an example of region 1 type field-aligned currents obtained by a 3D simulation for the magnetosphere-ionosphere coupling. Figure 3.17(a) is the field-aligned current at the ionospheric height caused by a twin-vortex convective flow at the

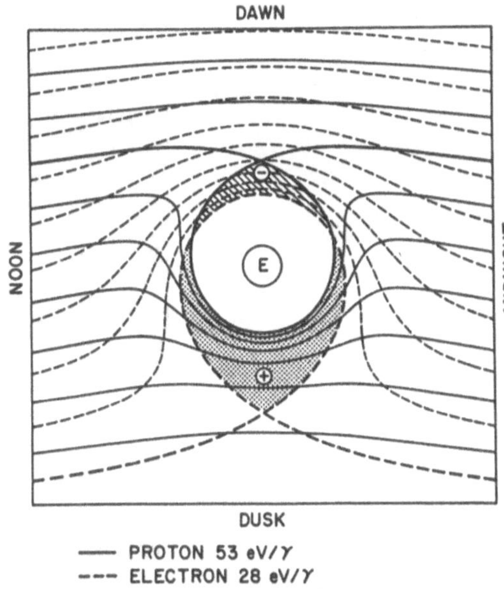

Fig. 3.16. Typical proton and electron orbits

magnetospheric equator and Fig. 3.17(b) is the electric potential projected from the magnetosphere to the ionosphere via Alfvén waves. The field-aligned current distribution along the field lines at a latitudinal cross-section of 0600 LT (dawn) is shown in Fig. 3.17(c).

Ring Current

As we have seen in Fig. 3.11, a dawn-dusk tail current can be generated to form a nearly anti-parallel field configuration in the distant tail unless the magnetosphere is empty of plasma. The dawn-dusk tail current is closed with semi-circular Chapman-Ferraro currents flowing on the magnetopause in the tail. The tail current is the diamagnetic current due to hot plasma sheet plasma and the Chapman-Ferraro current is the inertial current and/or the diamagnetic current of the solar wind plasma. Closer to the earth ($r < 7\,R_E$), the tail current will disappear in accordance with the disappearance of the plasma sheet. In its place the ∇B-drift current appears as a result of the difference in the directions of the electron and proton drift orbits.

The ∇B-drift current carried by particles with open orbits (Fig. 3.16) cannot be divergence-free by itself. Accordingly, a field-aligned current is generated to compensate for charge-neutrality. (Note that the divergence of the magnetization current is zero, hence the magnetization current cannot contribute to the field aligned

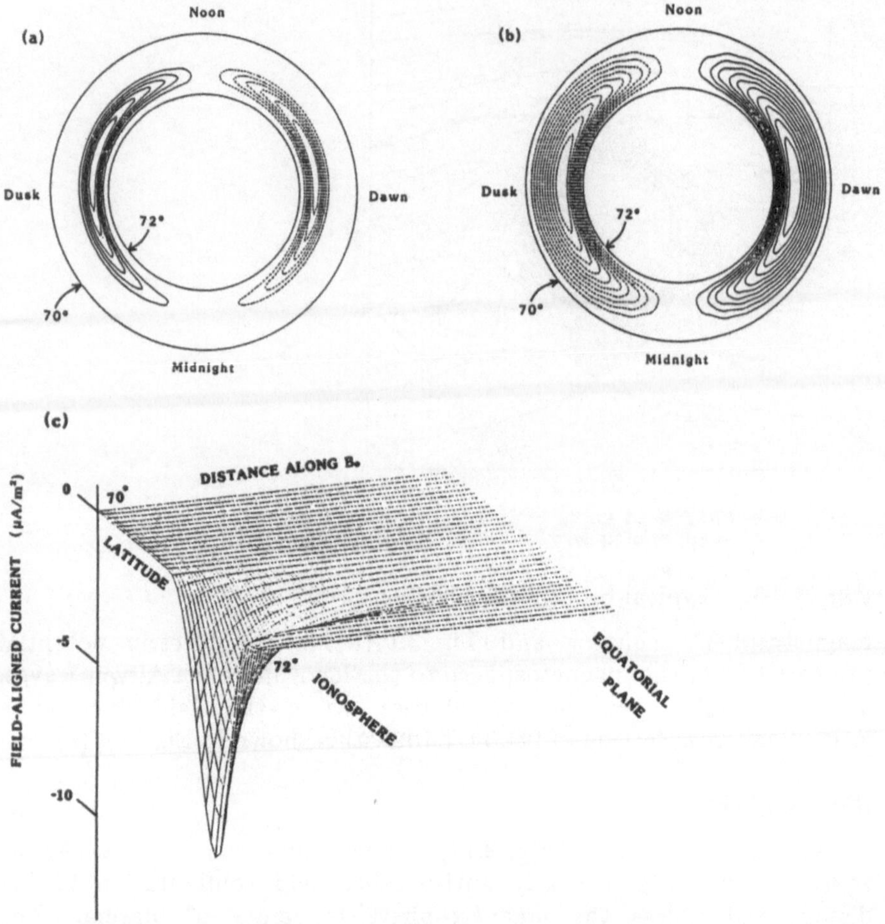

Fig. 3.17. Region 1 type field-aligned currents in the ionosphere **a** driven by a twin-vortex motion in the magnetospheric equator. Corresponding electric potential distribution is shown in **b**. **c** is the field-aligned current distribution along the geomagnetic field: A result of a computer simulation of the 3D magnetosphere - ionosphere coupling

current.) This charge-neutralizing field-aligned current is attributed to region 2 current. Such a drift current may contribute to a partial ring current since it is not closed by itself.

The ∇B-drift current carried by particles with closed orbits, on the other hand, can close by itself. In this case, the total current density may be evaluated by adding the magnetization current (as well as the curvature current). Then, as shown in Section 1.8, the ∇B current cancels with a part of the magnetization current and the resultant current is given by the diamagnetic current, $J_D = -\nabla p \times \hat{b}/B$. When hot plasma from the tail is injected into the magnetosphere, it is expected to be confined within a certain radius of the magnetosphere, hence ∇p is negative at the outer edge while it is positive in the inner edge of the injected plasma. Hence the diamagnetic current is directed westward at the outer edge while eastward at the inner edge. However, since the earth magnetic field is proportional to r^{-3}, the outer edge (westward) current dominates. Such a current can persist as long as the particles are free of pitch angle scattering. The current is called "ring current".

3.5 Ionosphere

We now consider the plasma of earth origin produced by photoionization of the solar radiation at the top of the earth's atmosphere.

Formation of Ionosphere

Let us first study how plasma of earth origin, that is, the "ionosphere", can be formed. Needless to say, the earth's atmosphere is filled with neutral atoms and molecules. These particles are primarily heated by the heat of the earth's solid surface which absorbs the solar radiation. The upward expansion of the heated atmospheric gas is pulled back to the earth by the earth's gravity. The (diffusive) balance is described by

$$\frac{\partial p}{\partial z} = -N\overline{m}g, \tag{3.5.1}$$

where p, N and $\overline{m}$ are the pressure, the average density and the average mass of the atmospheric gas, respectively; g is the earth's gravitational acceleration; z is the height above the earth's surface. In this equation we are only concerned with the vicinity of the earth's surface, say $0 \lesssim z \lesssim 10^3$ km, where g can be approximated by a constant value ($g = 9.8$ m/sec^2).

Assuming a constant temperature, Eq. (3.5.1) is solved to give

$$N = N_o \exp\left(-\frac{z}{H}\right), \tag{3.5.2}$$

where $H = \kappa T/\overline{m}g$, κ being the Boltzmann constant (1.38×10^{-23} joule/K). This indicates that the atmospheric gas density decreases exponentially with a scale height of H. For $T = 300$ K and $\overline{m} = 4.5 \times 10^{-26}$ kg, the scale height H becomes approximately 9 km.

Solar radiation in the form of photons can penetrate the earth's atmosphere without being blocked by the geomagnetic field. It is expected therefore that the radiation can interact directly with the atmosphere. The solar radiation spectrum can be described by Planck's radiation formula with a black body temperature of 6,000 K

$$I_\nu = \frac{2h\nu^3}{c^2} \frac{1}{\exp(h\nu/\kappa T) - 1}, \tag{3.5.3}$$

where h is Planck's constant (6.6×10^{-34} joule sec), ν is the frequency of the radiation and c is the speed of light. The spectrum for $T = 6{,}000$K is shown in Fig. 3.18. As can be seen from this figure, the spectrum peaks around 1 eV and includes considerable energy flux at 20 eV. Since the ionization potential of the atmospheric gas is known to be roughly 10 eV, (see Table 3.1), solar photon radiation can also ionize the atmospheric gas.

Let us assume a vertically stratified atmosphere, as is shown in Fig. 3.19. The solar radiation is incident on the atmosphere at an angle χ from the zenith. Then, the photon flux, dF, absorbed at a layer of dz is expressed by

$$dF = AFN \sec \chi \, dz, \tag{3.5.4}$$

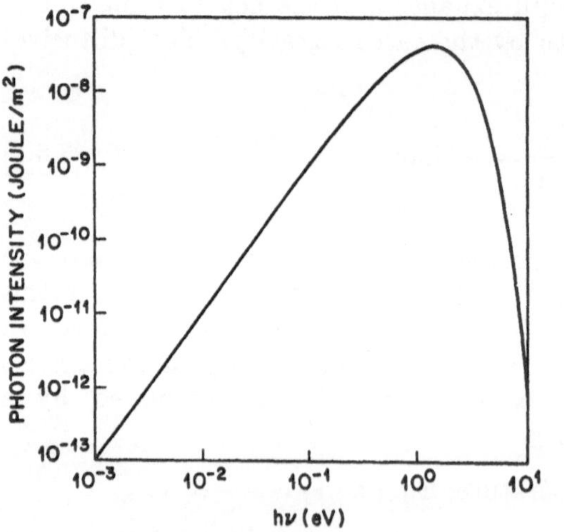

Fig. 3.18. Black body radiation spectrum corresponding to 6,000 K

where A [m^2] is the absorption coefficient (cross-section) for ionization and F [photon/m^2 sec] is the photon flux at $z = \infty$. The production rate of electrons q [electron/m^3 sec] is given by

$$q = \cos\chi \frac{dF}{dz} = AFN. \tag{3.5.5}$$

Here we assume that one photon can produce one electron-ion pair.

Integration of Eq. (3.5.4) yields

$$F = F_\infty \exp\left(-A \sec\chi \int_z^\infty N dz\right), \tag{3.5.6}$$

where F_∞ is the solar radiation flux far from the earth, $z = \infty$. Substituting Eq. (3.5.2) into Eq. (3.5.6) and then using Eq. (3.5.5), we obtain

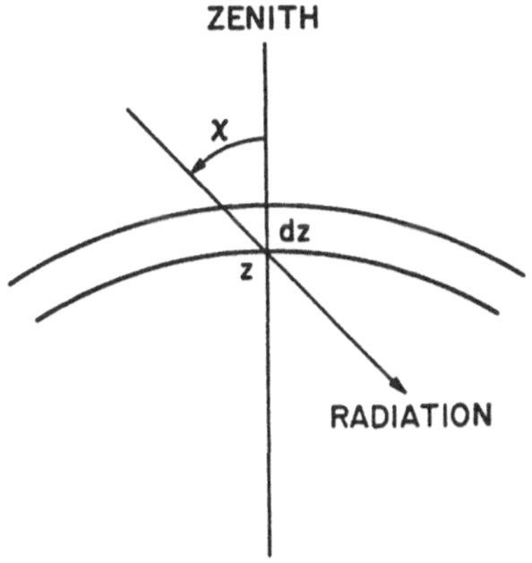

Fig. 3.19. Geometry of the direction of the solar radiation with respect to a vertically stratified atmosphere

Table 3.1 Ionization potential

Species	Ionization potential (eV)
H	13.60
O_2	12.20
O	13.62
N	14.53
N_2	15.58
NO	9.25

above 3,000 km neutral atoms and molecules almost vanish and only an ionized gas exists. The electron density is roughly $10^9 m^{-3}$ at 3,000 km and the ion constituents are almost all protons.

Conductivities and Currents

Figure 3.20 shows that the ionization rate is only of the order of 10^{-6} in the E layer ($z = 100 \sim 140$ km) and $10^{-2} \sim 10^{-5}$ in the F layer ($z = 200 \sim 500$ km). This means that the ionosphere is a partially ionized plasma. In contrast to the solar wind plasma and the magnetospheric plasma, which are really collisionless, the behavior of the ionospheric plasma is governed by collisions between charged particles and neutral particles.

The height distributions of collision frequencies and cyclotron frequencies are shown in Fig. 3.21, where ω_{ce} and ω_{ci} are electron and ion cyclotron frequencies, respectively, and ν_{en}, ν_{in} and ν_{ei} are electron-neutral, ion-neutral and electron-ion collision frequencies, respectively. From this figure it is seen that $\omega_{ce} >> \nu_{en}, \nu_{ei}$, indicating that the electron motion is almost collisionless. In contrast, $\nu_{in} >> \omega_{ci}$ in the E region and $\nu_{ei} \sim \omega_{ci}$ in the F region, so that the ion motion is collision-dominated.

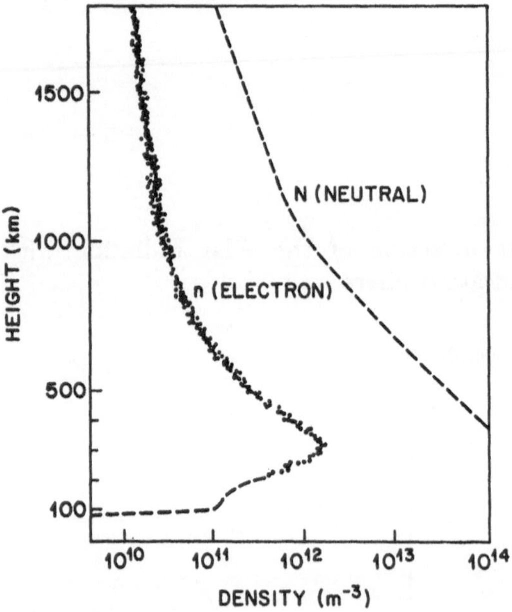

Fig. 3.20. Dayside electron density distribution measured by a rocket experiment (H. Oya and T. Obayashi) and neutral density distribution

$$q = AN_oF_\infty \exp\left[-\frac{z}{H} - AN_oH\sec\chi \exp\left(-\frac{z}{H}\right)\right]. \qquad (3.5.7)$$

This equation can be rewritten as

$$q = q_m \exp\left[1 - \tilde{z} - \exp(-\tilde{z})\right], \qquad (3.5.8)$$

where

$$\tilde{z} = (z - z_m)/H$$

$$z_m = H\ell n(AN_oH\sec\chi)$$

$$q_m = F_\infty \cos\chi / H\exp(1).$$

From Eq. (3.5.8) it is easily seen that the production rate is maximized at $z = z_m$ and the maximum rate is given by $q = q_m$. This is the Chapman model of ionosphere formation.

Since the production rate q of electrons is given, the electron density n is determined by knowing the loss rate L given as a function of n. In the lower ionosphere around 100 km, called the "E layer", recombination is the dominant loss process, namely,

$$L = \alpha n^2,$$

where α is the recombination coefficient. Thus $n = \sqrt{q/\alpha}$. In the higher ionosphere around 250 km, called the "F layer", electron attachment becomes dominant. Hence, the loss rate is given by

$$L = \beta n,$$

where β is the attachment coefficient. Thus $n = q/\beta$.

From the above arguments the electron density distribution can be explicitly obtained if the absorption coefficient A, the atmospheric gas density at the earth's surface N_o, the scale height H, the recombination coefficient α and the attachment coefficient β are known. Figure 3.20 shows an example of the daytime electron density distribution observed by a rocket experiment. The dotted line shows the atmospheric gas density.

From this figure it is seen that the electron density is maximized at a height of about 300 km where the density is roughly 10^{12} m^{-3}. Another small peak appears at about 100 km height where the density is roughly $10^{11} m^{-3}$. The former large layer corresponds to the F layer and the small layer to the E layer. Though not shown in this figure,

The equations of motion of electrons (e) and ions (i) are expressed by

$$n_e m_e \frac{d\mathbf{v}_e}{dt} + n_e m_e \nu_{en}(\mathbf{v}_e - \mathbf{V}_n) + n_e m_e \nu_{ei}(\mathbf{v}_e - \mathbf{v}_i)$$

$$= - n_e e(\mathbf{E} + \mathbf{v}_e \times \mathbf{E}) - \nabla p_e \qquad (3.5.9)$$

$$n_i m_i \frac{d\mathbf{v}_i}{dt} + n_i m_i \nu_{in}(\mathbf{v}_i - \mathbf{V}_n) + n_i m_i \nu_{ie}(\mathbf{v}_i - \mathbf{v}_e)$$

$$= n_i e(\mathbf{E} + \mathbf{v}_i \times \mathbf{B}) - \nabla p_i\,, \qquad (3.5.10)$$

where $\mathbf{V}_n$ is the velocity of neutral particles. In the ionosphere the plasma frequency ω_{pe} is $10^7 \sec^{-1}$ at 100 km height and $6 \times 10^7 \sec^{-1}$ at 250 km height. For stationary processes, we can assume charge neutrality, $n_e \simeq n_i \equiv n$, and also neglect the inertia terms.

Equations (3.5.9) and (3.5.10) are then reduced to

$$n\mathbf{v}_{\perp e} = \frac{\mathbf{A}_{\perp e}}{1 + (\omega_{ce}\tau_e)^2} - \frac{\omega_{ce}\tau_e}{1 + (\omega_{ce}\tau_e)^2}\mathbf{A}_{\perp e} \times \mathbf{b} \qquad (3.5.11)$$

$$n\mathbf{v}_{\perp i} = \frac{\mathbf{A}_{\perp i}}{1 + (\omega_{ci}\tau_i)^2} + \frac{\omega_{ci}\tau_i}{1 + (\omega_{ci}\tau_i)^2}\mathbf{A}_{\perp i} \times \mathbf{b} \qquad (3.5.12)$$

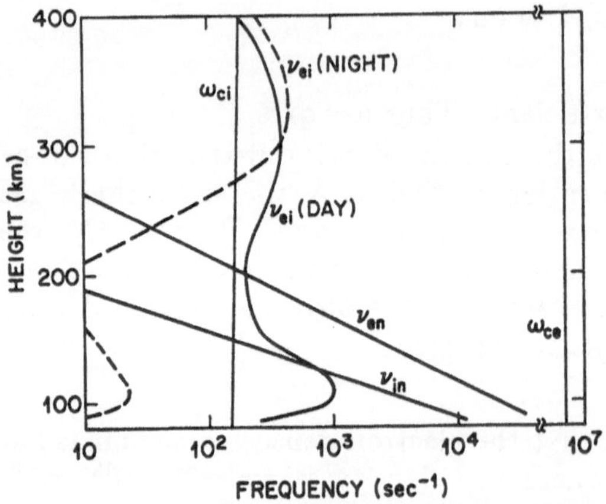

Fig. 3.21.

Fig. 3.21. Height distributions of collision frequencies and cyclotron frequencies in the ionosphere

$$nv_{\parallel e} = -\frac{ne}{m_e\nu_e}E_\parallel - D_e\nabla_\parallel n$$

$$+\frac{\nu_{en}}{\nu_e}nV_{\parallel n} + \frac{\nu_{ei}}{\nu_e}nv_{\parallel i} \tag{3.5.13}$$

$$nv_{\parallel i} = \frac{ne}{m_i\nu_i}E_\parallel - \quad D_e\nabla_\parallel n$$

$$+\frac{\nu_{in}}{\nu_i}nV_{\parallel n} + \frac{\nu_{ie}}{\nu_i}nv_{\parallel e}\,, \tag{3.5.14}$$

where $\nu_e = \nu_{en} + \nu_{ei}$, $\nu_i = \nu_{in} + \nu_{ie}$, $\tau_e = \nu_e^{-1}$, $\tau_i = \nu_i^{-1}$, $D_e = T_e/m_e\nu_e$, and $D_i = T_i/m_i\nu_i$;

$$\mathbf{A}_{\perp e} = -n(\omega_{ce}\tau_e)\frac{\mathbf{E}_\perp}{B} - D_e\nabla_\perp n + \frac{\nu_{en}}{\nu_e}n\mathbf{V}_{\perp n} + \frac{\nu_{ei}}{\nu_e}n\mathbf{v}_{\perp i} \tag{3.5.15}$$

$$\mathbf{A}_{\perp i} = n(\omega_{ci}\tau_i)\frac{\mathbf{E}_\perp}{B} - D_i\nabla_\perp n + \frac{\nu_{in}}{\nu_i}n\mathbf{V}_{\perp n} + \frac{\nu_{ie}}{\nu_i}n\mathbf{v}_{\perp e}\,. \tag{3.5.16}$$

The important feature to be deduced from these equations is that electrons and ions behave very differently under the action of an electric field because of the large difference in $(\omega_{ce}\tau_e)$ and $(\omega_{ci}\tau_i)$. In a collisionless plasma where $\omega_{ce}\tau_e >> 1$ and $\omega_{ci}\tau_i >> 1$, Eqs. (3.5.11) and (3.5.12) reduce to $\mathbf{v}_{\perp e} = \mathbf{v}_{\perp i} = \mathbf{E}\times\mathbf{B}/B^2$, thus, no current will flow perpendicular to the magnetic field in the presence of the perpendicular electric field. This implies that a collisionless plasma is an insulator in the direction perpendicular to the magnetic field, although it is a good conductor parallel to it. In contrast, in the ionosphere a dc current can easily flow across the magnetic field in response to a perpendicular electric field. This fact leads to an important issue in the magnetosphere-ionosphere interaction, namely, that a field-aligned current of magnetospheric origin can close in the ionosphere. In other words, dc field-aligned currents of magnetospheric origin can flow because of the existence of the ionosphere.

As we have seen in the foregoing section, an electric field and current can be produced in a direction perpendicular to the magnetic field in the magnetosphere. If the whole magnetosphere were filled with collisionless (dissipationless) plasmas, the magnetic and plasma energies transported from the solar wind would be stored persistently in the magnetosphere. However, since the magnetic field lines are connected to the collisional ionosphere, the magnetospheric electric field mapped down to the ionosphere along the magnetic field can give rise to an electric current across the magnetic field in the ionosphere, whereby the stored magnetospheric energy can be dissipated ohmically.

Let us study this situation in a more quantitative way. For simplicity but without loss of generality, we shall neglect the neutral wind and the diffusion terms. Furthermore, we direct our attention to the E region ionosphere where the ν_{ei} term can be omitted. Then Eqs. (3.5.11) through (3.5.14) are reduced to

$$\mathbf{v}_{\perp e} = -\mu_{Pe}\mathbf{E}_{\perp} + \mu_{He}\frac{\mathbf{E}\times\mathbf{B}}{B} \tag{3.5.17}$$

$$\mathbf{v}_{\perp i} = \mu_{Pi}\mathbf{E}_{\perp} + \mu_{Hi}\frac{\mathbf{E}\times\mathbf{B}}{B} \tag{3.5.18}$$

$$v_{\parallel e} = -\mu_{\parallel e}E_{\parallel} \tag{3.5.19}$$

$$v_{\parallel i} = \mu_{\parallel i}E_{\parallel}, \tag{3.5.20}$$

where

$$\mu_{\parallel e} = \frac{e}{m_e\nu_e}$$

$$\mu_{Pe} = \frac{\mu_{\parallel e}}{1+(\omega_{ce}\tau_e)^2}$$

$$\mu_{He} = \frac{\mu_{\parallel e}(\omega_{ce}\tau_e)}{1+(\omega_{ce}\tau_e)^2}$$

$$\mu_{\parallel i} = \frac{e}{m_i\nu_i}$$

$$\mu_{Pi} = \frac{\mu_{\parallel i}}{1+(\omega_{ci}\tau_i)^2}$$

$$\mu_{Hi} = \frac{\mu_{\parallel i}(\omega_{ci}\tau_i)}{1+(\omega_{ci}\tau_i)^2}.$$

Here $\mu_{\parallel}$, μ_P and μ_H are called parallel, Pedersen and Hall mobilities, respectively.

These equations are combined to give the ionospheric current J_I as

$$J_{\perp I} = \sigma_P \mathbf{E}_\perp - \sigma_H \frac{\mathbf{E}_\perp \times \mathbf{B}}{B} \quad (3.5.21)$$

$$J_{\parallel I} = \sigma_\parallel E_\parallel , \quad (3.5.22)$$

where

$$\sigma_P = ne(\mu_{Pe} + \mu_{Pi})$$

$$\sigma_H = ne(\mu_{He} - \mu_{He})$$

$$\sigma_\parallel = ne(\mu_{\parallel i} + \mu_{\parallel e}).$$

Here σ_P, σ_H and $\sigma_\parallel$ are called Pedersen, Hall and parallel conductivities, respectively.

The height distributions of the conductivities are shown in Fig. 3.22. From this figure it is seen that $\sigma_\parallel > \sigma_H > \sigma_P$ in the E region ionosphere. Because of the high parallel conductivity the parallel potential is usually short-circuited, so that we can reasonably assume $E_\parallel = 0$. More importantly, it is seen that the E region ionosphere has a rather high conductivity perpendicular to the magnetic field. When space charges appear across the magnetic field in the magnetosphere, therefore, they can discharge along the magnetic field lines and through the E region ionosphere.

From Eq. (3.5.21) the ohmic dissipation in the ionosphere is given by

$$\int_I \mathbf{J}_{\perp I} \cdot \mathbf{E}_\perp dV = \int_I \sigma_P E_\perp^2 \, dV . \quad (3.5.23)$$

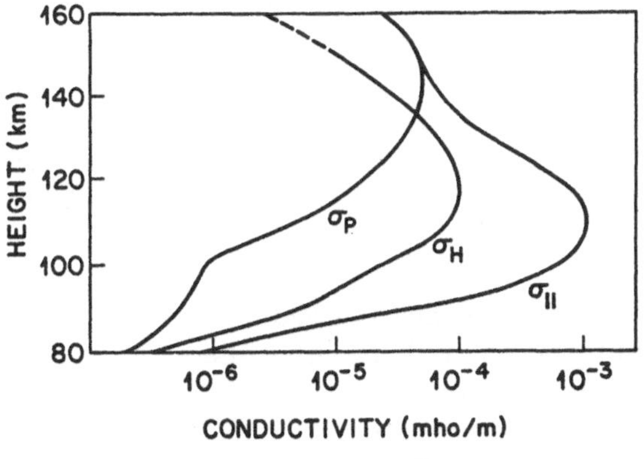

Fig. 3.22. Height distributions of ionosphere conductivities

Obviously the Hall conductivity does not explicitly contribute to dissipation. In a homogeneous ionosphere where the Hall conductivity is constant, the Hall current itself is divergence-free, so that it plays only a passive role in the ionospheric dynamics. In an inhomogeneous ionosphere, however, the Hall term can exhibit an active role, as we will see in the following.

Let us take a rectangular coordinate system (x, y, z) in which a uniform magnetic field B points to the positive z axis and a uniform electric field $\mathbf{E}_o = (E_{xo}, E_{yo}, O)$ is externally given. The external electric field can be a neutral wind driven field, e.g., the so-called ionospheric dynamo field, or an electric field of magnetospheric origin in the high latitude ionosphere. Suppose that the Pedersen and Hall conductivities have discontinuities across a certain layer, i.e., $-L/2 \leq y \leq L/2$, from σ_{Po} and σ_{Ho} to σ_{Pl} and σ_{Hl} $(\sigma_{Pl} > \sigma_{Po}, \sigma_{Hl} > \sigma_{HO})$. Then, from the current continuity across the high conductive layer (y direction), a polarization electric field ΔE_y appears in the high conductive layer, which is given by

$$\Delta E_y = E_{yl} - E_{yo} = -\frac{\sigma_{Pl} - \sigma_{Po}}{\sigma_{Pl}} E_{yo} - \frac{\sigma_{Hl} - \sigma_{HO}}{\sigma_{Pl}} E_{xo} . \quad (3.5.24)$$

When $\sigma_{Pl} >> \sigma_{Po}$ and $\sigma_{Hl} >> \sigma_{Ho}$, we have $\Delta E_y \approx -E_{yo} - (\sigma_{Hl}/\sigma_{Pl}) E_{xo}$. This indicates that the Pedersen component is completely shielded in the high conductive layer but the Hall term generates an additional field proportional to E_{xo}.

In the equatorial ionosphere the magnetic field is horizontal and points to the north. As can be seen from the height distributions of the conductivities shown in Fig. 3.22, the conductivities, particularly the Hall conductivity, are peaked in the E layer. Therefore, the above argument can be applied when we assign the x axis west, the y axis upward and the z axis to the north. The neutral wind-driven dynamo electric field in the equatorial ionosphere usually points westwards (or eastwards). Assuming $\sigma_{P1} >> \sigma_{Po}$ and $\sigma_{H1} >> \sigma_{Ho}$, therefore, we obtain

$$\Delta E_y = -\frac{\sigma_H}{\sigma_P} E_{xo} , \quad (3.5.25)$$

where the subscript 1 representing the highly conductive E region is omitted and the same abbreviation will be used hereafter. The westward electric current is thus given by

$$J_x = \sigma_P [1 + (\frac{\sigma_H}{\sigma_P})^2] E_{xo} . \quad (3.5.26)$$

This indicates that the westward current at the equator is intensified by $[1+(\sigma_H/\sigma_P)^2]$ owing to the vertical polarization effect. In the E region $\sigma_H/\sigma_P \approx 10$, as can be seen in Fig. 3.22. Thus, the current is expected to be intensified by a hundred times. In fact, such an unexpectedly large westward electric current is observed in the daytime equatorial ionosphere. This is called the "equatorial electrojet". The effective conductivity $\sigma_P[1+(\sigma_H/\sigma_P)^2]$ is called the Cowling conductivity.

In the polar ionosphere where the magnetic field is almost vertical and extends to the magnetosphere, a similar situation can be realized. Auroral arcs are usually elongated in the east-west direction and very narrow in the north-south direction. Furthermore, the electron density is expected to be greatly intensified in an auroral arc. When we assign the x axis to the west, the y axis to the south and the z axis upward, therefore, the same argument can be employed. Unlike the equatorial case, however, the magnetic field extends to a huge volume as the magnetosphere. Therefore, it is expected that the polarized space charges generated at the conductivity discontinuities can escape to the magnetosphere. In order to attack this problem rigorously, we must solve self-consistently the magnetosphere-ionosphere coupling. We leave the dynamical aspect of this process to Volume II. But it should be emphasized that polarization electric fields, Eq. (3.5.24) or Eq. (3.5.25), arising from the density inhomogeneities in the polar ionosphere can drive field-aligned currents of ionospheric origin. In the closed field line region, however, the field-aligned current originating from the polar ionosphere cannot flow steadily because the magnetosphere behaves as an insulator perpendicular to the magnetic field; hence, a polarization effect might be expected. In the open field line region, on the other hand, the space charges generated at the conductivity discontinuities will continuously escape from the ionosphere because the magnetospheric volume is indefinite, hence, the magnetospheric capacitance is infinite. Thus, polarization effects might not operate. In this respect, it is important to note that whether the field lines are closed or open, when they are time-dependent, field-aligned currents of ionospheric origin can flow into the magnetosphere. Thus, a feedback coupling is expected in this case. This will be discussed in detail in Vol. II.

3.6 Magnetosphere-Ionosphere Coupling

Both in the magnetosphere and in the ionosphere plasma currents and electric fields can be generated perpendicular to the magnetic field. The field lines extend to the magnetosphere, threading the high latitude ionosphere in one hemisphere, and return to the high latitude ionosphere in the other hemisphere. Furthermore, charged particles, particularly electrons, can freely move along them. These facts indicate that the magnetosphere and the ionosphere are strongly and electrodynamically coupled. In the following discussion we formulate a stationary coupling through field-aligned currents between the magnetosphere and the ionosphere.

In general the dimensions of the magnetosphere are enormously large compared to the ionosphere. Therefore information on the magnetosphere cannot instantly be communicated to the ionosphere. For phenomena accompanying field-aligned currents Alfveń waves play the role of carrying information (currents) between the magnetosphere and the ionosphere. Since it takes a few minutes for an Alfveń wave to propagate between the magnetospheric equator and the ionosphere, magnetospheric phenomena of time scales longer than one minute may strongly be influenced by the ionosphere, and vice versa.

Since we are concerned here with stationary problems, we can consider that Alfveń waves have already traveled many times back and forth between the magnetosphere and the ionosphere and that a stationary field-aligned current system is established. Thus, the stationary magnetosphere-ionosphere coupling can be formulated as follows:

First, we evaluate the electric currents perpendicular to the magnetic field both in the magnetosphere and in the ionosphere. Then, we take their divergence, which should be equal to minus the divergence of the field-aligned currents. We integrate them along the field lines over the whole magnetosphere and the ionosphere, respectively, to obtain the net field-aligned currents at the top of the ionosphere (bottom of the magnetosphere). Finally, we equate them to assure the continuity of the current.

From Eq. (3.5.21) the ionospheric part of the upward field-aligned current is given by

$$J_{\|I} = -\int_I \nabla_\perp \cdot (\sigma_P \mathbf{E}_I - \sigma_H \frac{\mathbf{E}_I \times \mathbf{B}_I}{B_I}) dz . \qquad (3.6.1)$$

Here the parallel current is assumed to vanish at the bottom of the ionosphere. This equation can be approximated by

$$J_{\|I} \cong -\nabla_\perp \cdot (\Sigma_P \mathbf{E}_I - \Sigma_H \frac{\mathbf{E}_I \times \mathbf{B}_I}{B_I})$$

$$\cong -\nabla_\perp \cdot \Sigma_P \mathbf{E}_I + \frac{\mathbf{E}_I \times \mathbf{B}_I}{B_I} \cdot \nabla \Sigma_H , \tag{3.6.2}$$

where $\Sigma_P = h\sigma_P$ (height-integrated Pedersen conductivity) and $\Sigma_H = h\sigma_H$ (height-integrated Hall conductivity), and h is the thickness of the E layer.

The magnetospheric current, on the other hand, is given by Eq. (1.6.10), namely,

$$J_M = \frac{\mathbf{B} \times \nabla p}{B^2} - \frac{nm}{B^2} \frac{d\mathbf{v}}{dt} \times \mathbf{B} . \tag{3.6.3}$$

Taking the divergence of Eq. (3.6.3) yields

$$\nabla_\perp \cdot \mathbf{J}_M = \nabla_\perp \cdot (p \nabla \times \frac{\mathbf{B}}{B^2}) - nm \frac{d}{dt}(\frac{\Omega}{B})$$

$$- \mathbf{J}_{in} \cdot \nabla \ell n\, n , \tag{3.6.4}$$

where $\Omega = \mathbf{b} \cdot \nabla \times \mathbf{v}$ (vorticity) and $\mathbf{J}_{in}$ (inertial current) is the second term on the right hand side of Eq. (3.6.3). After some manipulation, Eq. (3.6.4) is rewritten as

$$\nabla_\perp \cdot \mathbf{J}_M = 2\mathbf{b} \cdot [\nabla p \times \nabla (\frac{1}{B})] - nm \frac{d}{dt}(\frac{\Omega}{B})$$

$$- J_{in} \cdot \nabla \ell n\, nB^2 . \tag{3.6.5}$$

Since

$$\nabla_\perp \cdot J_M = -\nabla_\parallel \cdot J_\parallel = -B \frac{\partial}{\partial z}(\frac{J_\parallel}{B}), \tag{3.6.6}$$

we have

$$[\frac{J_\parallel}{B}]_{z_1}^{z_2} = \int_{z_1}^{z_2} \frac{1}{B} \{-2\mathbf{b} \cdot [\nabla p \times \nabla (\frac{1}{B})] + nm \frac{d}{dt}(\frac{\Omega}{B})$$

$$+ \mathbf{J}_{in} \cdot \nabla \ell n\, nB^2\} dz , \tag{3.6.7}$$

where z_1 represents the top of one hemisphere and z_2 the top of the other hemisphere. For the sake of simplicity we assume symmetry between the northern and southern hemispheres and that no inertial current is present in the background. Then, Eq. (3.6.7) reduces to

$$\frac{J_{\|\mathrm{I}}}{B_\mathrm{I}} = \frac{1}{2}\int\{\mathbf{b}\cdot[\nabla p\times\nabla(\frac{1}{B^2})] - \frac{nm}{B^2}\frac{d\Omega}{dt}\}dz\,, \tag{3.6.8}$$

where the integral is over the entire field line concerned.

The first term on the right hand side of Eq. (3.6.8) represents the divergence of the magnetospheric current and the second the discharge of the magnetospheric charge (vorticity). The first term can be attributed to the divergence of the ring current due to the difference in drift orbits of electrons and protons discussed in Section 3.4. In this case the pressure can be calculated when a particle distribution is given as a function of the magnetic moment. The second term can be attributed to the discharge of magnetospheric space charges due to vortex motion. When the vorticity is directly supplied from the drag of the solar wind, the discharge must be balanced by the generation of vorticity in the steady state. Denoting the vorticity generation rate by S, we have

$$\mathbf{v}\cdot\nabla\Omega = S\,. \tag{3.6.9}$$

Combining Eqs. (3.6.2) and (3.6.8), we obtain

$$\nabla_\perp\cdot\Sigma_p\nabla V + \frac{\mathbf{B}_\mathrm{I}\times\nabla V}{B_\mathrm{I}}\cdot\nabla\Sigma_H$$

$$= \frac{1}{2}\int\{\mathbf{b}\cdot[\nabla p\times\nabla(\frac{1}{B^2})] - \frac{nm}{B^2}S\}dz\,. \tag{3.6.10}$$

In this equation we have used an electric potential V instead of the ionospheric electric field. This is because in a stationary state the potential is expected to be constant along the magnetic field. Now that the vorticity in Eq. (3.6.9) is related to the electric potential and that the pressure can be calculated when particle orbits are known in the presence of the electric field, namely, the electric potential [see Eq. (3.4.62)], Eq. (3.6.10) is in principle solvable. Figure 3.17 shown previously is one solution of Eq. (3.6.8) for a case where a twin-vortex pattern of the plasma flow is given in the magnetosphere, the plasma is assumed to be cold and the ionospheric density has a day-night asymmetry.

When a field-aligned current exists, a closure ionospheric current is required, consequently, the ionosphere is ohmically heated, see Eq. (3.5.23). If we can exclude the contribution of the neutral wind dynamo in the ionosphere, the primary energy source of the ohmic dissipation in the ionosphere must be in the magnetosphere. As we have seen above, there are two potential candidates for an energy source in the magnetosphere, namely, the diamagnetic current and the vortex motion.

From Eq. (3.6.3) we have

$$\int_M \mathrm{J_M} \cdot \mathbf{E} dV = \int \mathbf{v} \cdot (\nabla p + nm\frac{d\mathbf{v}}{dt}) dV = \int \mathbf{v} \cdot (\mathrm{J_M} \times \mathbf{B}) dV . \quad (3.6.11)$$

This equation implies that when the Ampere force has a component parallel to the electric drift, the magnetosphere will behave as an energy absorber. On the other hand, when there is an antiparallel component, the magnetosphere will act as an energy supplier to the ionosphere. This means that when the diamagnetic current is the source of the field-aligned current (region 2 current), the pressure force must oppose the electric drift, that is

$$\int_M \mathbf{v} \cdot \nabla p dV < 0 . \quad (3.6.12)$$

In this case the magnetospheric thermal energy is the primary energy source. When magnetospheric convection is the source, the inertial force must oppose to the electric drift, namely,

$$\int_M nm\mathbf{v} \cdot \frac{d\mathbf{v}}{dt} dV = \frac{1}{2}\int nm\frac{d}{dt} v^2 dV < 0 . \quad (3.6.13)$$

In this case the magnetospheric flow energy is the primary energy source.

From the energy balance between the magnetosphere and the ionosphere, i.e., from Eqs. (3.5.23) and (3.6.11), we have

$$\int_I \sigma_P E_\perp^2 dV + \int_M \mathbf{v} \cdot (\nabla p + nm\frac{d\mathbf{v}}{dt}) dV = 0 . \quad (3.6.14)$$

In order to have a stationary field-aligned current system, therefore, we must have a hot particle source and/or a plasma flow source in the magnetosphere which can be steadily supplied from the solar wind.

References for Chapter 3

W. I. Axford, *The Interaction of the Solar Wind with the Interstellar Medium, In: Solar Wind,* edited by C. P. Sonett, P. J. Coleman, Jr., and J. M. Wilcox, National Aeronautics and Space Administration, Washington, D.C., pp. 609-660, 1972.

J. B. Brandt, *Introduction to the Solar Wind,* W. H. Freeman and Company, San Francisco, 1970.

W. D. Hayes and R. F. Probstein, *Hypersonic Flow Theory,* Academic Press, New York, London, 1959.

A. J. Hundhausen, *Coronal Expansion and Solar Wind,* Springer-Verlag, New York, Heidelberg, Berlin, 1972.

L. D. Landau and E. M. Lifshitz, *Electrodynamics of Continuous Media,* Pergamon Press, Oxford, New York, Toronto, Sydney, Paris, Braunschweig, 1975.

A. Nishida, *Geomagnetic Diagnosis of the Magnetosphere,* Springer-Verlag, New York, Heidelberg, Berlin, 1978.

T. Obayashi, *Space Physics* (in Japanese), Shokabo, Tokyo, 1970.

Subject Index